Handbook of **Quantitative Ecology**

HANDBOOK OF

QUANTITATIVE ECOLOGY

JUSTIN KITZES

THE UNIVERSITY OF CHICAGO PRESS

CHICAGO AND LONDON

The University of Chicago Press, Chicago 60637
The University of Chicago Press, Ltd., London
© 2022 by The University of Chicago
All rights reserved. No part of this book may be used or reproduced in any manner whatsoever without written permission, except in the case of brief quotations in critical articles and reviews. For more information, contact the University of Chicago Press, 1427 E. 60th St., Chicago, IL 60637.
Published 2022
Printed and bound by CPI Group (UK) Ltd, Croydon, CR0 4YY

31 30 29 28 27 26 25 24 23 22   1 2 3 4 5

ISBN-13: 978-0-226-81832-0 (cloth)
ISBN-13: 978-0-226-81832-0 (paper)
ISBN-13: 978-0-226-81833-7 (e-book)
DOI: https://doi.org/10.7208/chicago/9780226818337.001.0001

Library of Congress Cataloging-in-Publication Data

Names: Kitzes, Justin, 1982– author.
Title: Handbook of quantitative ecology / Justin Kitzes.
Description: Chicago ; London : The University of Chicago Press, 2022. | Includes bibliographical references and index.
Identifiers: LCCN 2021049434 | ISBN 9780226818320 (cloth) | ISBN 9780226818344 (paperback) | ISBN 9780226818337 (ebook)
Subjects: LCSH: Ecology—Mathematical models.
Classification: LCC QH541.15.M3 K58 2022 | DDC 577.01/51—dc23
LC record available at https://lccn.loc.gov/2021049434

♾ This paper meets the requirements of ANSI/NISO Z39.48-1992 (Permanence of Paper).

To Adina and John,

who helped me learn

both the how and

the why of these

methods

Contents

Introduction   1

PART I Change over Time
1   Introducing Difference Equations   7
2   Duckweed on a Pond: Exponential Growth   11
3   Throwing Shade I: Logistic Growth   16
4   Throwing Shade II: Lotka-Volterra Competition   22
5   Rabies Removal: SIR Models   28

PART II Understanding Uncertainty
6   Introducing Probability   35
7   A Bird in the Cam I: Single-Variable Probability   38
8   A Bird in the Cam II: Two-Variable Probability   43
9   Picking Ticks: Bayes's Rule   50
10   Rabbit Rates: Probability Distributions   57

PART III Modeling Multiple States
11   Introducing Matrix Models   65
12   Imagine All the Beetles: Age-Structured Models   69
13   The Road to Succession: Transition Matrices   76
14   A Pair of Populations: Absorption   81
15   Fish Finders: Diffusion   88

PART IV Explaining Data
16   Introducing Statistics   101
17   Seedling Counts I: Maximum Likelihood   103
18   Seedling Counts II: Model Selection   110
19   Flattened Frogs I: Generalized Linear Models   115
20   Flattened Frogs II: Hypothesis Testing   121

PART V Expanding the Toolbox

21 Other Techniques   131
22 Bird Islands: Graphical Thinking   132
23 Max Plant Institute: Optimization   140
24 Bears with Me: Stochastic Simulation   146
25 Natives in the Neighborhood: Cellular Automata   151

References   157
Index   159

Introduction

Given that you have picked up and started to read this book, I expect that you have some interest in the science of ecology. Particularly if you are a student, you may even be familiar with some of the many different subfields of ecology that have developed over the years: population ecology, community ecology, ecosystem ecology, behavioral ecology, organismal ecology, physiological ecology, landscape ecology, spatial ecology, macroecology, and so on. With all of these different "ecologies" to learn about, you might be wondering what a book on "quantitative ecology" is about, and why you might want to read it.

Put simply, **quantitative ecology** is about the application of mathematical (as in equations) and computational (as in computer programming) methods to the practice of ecology. These methods are important parts of many of the subfields of ecology listed above, and include techniques related to population modeling, working with probabilities, data analysis and statistics, and other numerical methods such as optimization and simulation. The purpose of this book is to provide an introductory guide, or a handbook, that explains the basic principles of these diverse and important quantitative methods in an accessible and intuitive manner.

I was inspired to write this book in response to two observations about ecology and ecologists. First, it is becoming nearly impossible to understand modern ecology without some foundational understanding of the quantitative methods used by modern ecologists. This will be obvious to professional ecologists who read peer-reviewed papers as part of their jobs, as well as to students reading any non-introductory ecology textbook. Second, I have found that many (though certainly not all) ecologists consider math and computing to be one of their weakest areas. I suspect that at least some of you reading this introduction feel similarly.

My fundamental belief in writing this book is that anyone with the ability to understand the non-quantitative aspects of ecology is also capable of understanding the basic principles of quantitative ecology. I also believe that, with the right materials, nearly all ecologists are capable of teaching themselves the basics of these quantitative methods more easily than they might think. The challenge is to provide an introduction to these methods that does not require much prior knowledge of math or computing, that shows clearly how these methods are *used* in ecological applications, and that explains clearly the "missing" steps that are often left out of textbooks. The goal of this book is to help you learn the foundations of quantitative ecology by directly addressing these three challenges to learning.

In terms of prior experience, I will assume that you have a background in ecology and biology equivalent to one or two introductory undergraduate classes. On the other hand, I will assume that you may have very little formal background in math or computing, or that you could use a refresher on the concepts that you learned at one point. To be very specific, *understanding the material in this book does not require any knowledge of math beyond algebra, and does not require any knowledge of programming beyond the ability to use a spreadsheet.* Somewhat unusually for a book on quantitative methods, this book does not require you to understand or use calculus. Every method that we cover without using calculus, however, has a direct conceptual analogue to a method that does involve calculus, and I point these out throughout the book for those who are interested.

This book also takes an explicitly bottom-up, rather than top-down, approach to learning each quantitative method. You are undoubtedly familiar with the top-down approach of most textbooks, which present an equation for population growth, for example, as a fact that seems to appear out of nowhere before applying that equation to a few examples. In contrast, we will take a bottom-up approach in which each chapter presents a specific ecological problem to be solved. In the process of taking a logical, step-by-step approach to solving that problem, we will arrive at a general principle, such as the exact same population growth equation. I believe that this bottom-up approach makes it much easier to understand *why* certain quantitative methods work the way they do. I also believe that understanding the basic intuition behind the methods, rather than every technical detail, is the best foundation for your future growth as a quantitative ecologist.

The book itself is broken down into five parts, each of which covers a broad group of quantitative methods: difference equations, probability, matrix models, statistics, and a final "grab bag" of techniques that didn't fit neatly in the other parts. Each part contains a short introduction followed by four chapters in the bottom-up style described above. Each of those chapters begins with a specific ecological problem, which is then solved over the course of the chapter. Given the breadth of concepts presented in this book, each chapter is able to present only an initial taste of the method that it covers. Each chapter then ends with a section titled "Next Steps," which provides suggestions for readers who wish to continue learning about a particular method or technique. The chapters within each of the five parts of the book generally build on one another, and you will probably want to at least skim through the earlier chapters in each part before tackling the later ones.

While there is no need to read the parts of this book in sequence, readers who consider their quantitative background to be fairly thin might want to proceed through the five parts in the order presented. This is because, unavoidably, some of the basic techniques presented in the earlier chapters are used in later chapters. In particular, the basic material on constructing and understanding equations that is presented in part 1, and the instructions for using Google Sheets, will be used throughout the book. Understanding matrix models (part 3) and basic statistics (part 4) also requires some understanding of difference equations (part 1) and probability (part 2). If you do choose to skip ahead and find that you do need a refresher, chapters in these later parts of the book refer back specifically to the earlier chapters that they build on.

I hope that readers with a more extensive quantitative background will still find material in this book that can help to fill gaps in their knowledge. For example, ecologists with a statistical inclination may benefit from the ideas behind simulating a null distribution (chap. 20) or the principles of optimization (chap. 23). Those with a background in population modeling may be interested in the similarities between multi-state population projection models and multivariate probability models (chap. 13) or in basic stochastic simulation methods (chap. 24). And everyone stands to benefit from a better understanding of Bayes's rule (chap. 9) if you're not already deeply familiar with this equation, just in case you or anyone you know is ever diagnosed with a rare disease.

Although I won't assume much in the way of prior background, I do hope that you will be willing to bring some time and effort to the material in each chapter. You should not expect to be able to read a chapter casually one time through and understand everything that's in it. Instead, I would encourage you to read each chapter slowly, thinking carefully about how the system being described actually works. Most importantly, *do not skip over the equations*. Every equation in this book is explained in words as well as symbols. Spending the time to puzzle through the meaning of each equation will be the key to understanding how and why certain methods work the way that they do, as well as to seeing how they form the foundation for more complex methods.

In addition to reading each equation carefully, I hope that you will take the time to perform the calculations needed to solve each problem. To help you in this, instructions are provided for performing all calculations in Google Sheets. You should be able to find instructions online for creating a free Google account, if you don't have one already, and for creating a new sheet to use for these calculations. Fully completed sheets for every chapter that uses them are also available at http://www.handbookofquantitativeecology.org. I strongly encourage you, however, to try out the calculations for yourself before looking at the answers provided there. You may be surprised by how much better you understand the calculations if you take the extra time to perform them yourself.

Last but not least, I have many people to thank for their roles in helping to get this book into your hands. I thank my colleagues and students Lauren Chronister, David Clark, Andrea Fetters, Vero Iriart, Cassie Olmsted, Tessa Rhinehart, Lauren Schricker, Daniel Turek, Taylor Zallek, and members of the Harte Lab at the University of California, Berkeley for providing input on drafts of the text. I also thank my former PhD and postdoctoral advisor, Dr. John Harte, whose books *Consider a Spherical Cow* and *Consider a Cylindrical Cow* provided the inspiration for the structure of this book. The Berkeley Institute for Data Science at the University of California, Berkeley, and the University of Pittsburgh Department of Biological Sciences both provided financial support during the writing of this book. Finally, I thank the editors and staff at the University of Chicago Press for their work in editing, producing, and ultimately distributing this book to you.

Part I **Change over Time**

1 Introducing Difference Equations

Some of the most basic questions in quantitative ecology ask how things will change over time. Commonly, these "things" are individual organisms that form a population of a species. We are often interested in projecting, for example, whether a population is expected to increase or decrease in size over time, and at what rate.

The most basic method for projecting change over time is a **difference equation**. The core idea behind a difference equation is that we can start with a count of things present right now and use an equation to predict how many things will be present in the next time period. Then, we can take the number of things present in that next time period and use the same equation to calculate how many will be present in the time period after that, and so on.

As an example, imagine that you are planting a row of seeds in a garden. You can plant 2 seeds per minute on average. How many seeds will you have planted after 10 minutes? It should be obvious that at a rate of 2 seeds per minute, you will be able to plant $2 \times 10 = 20$ seeds in 10 minutes.

Let's write this simple calculation down as a difference equation to see how this method works. To start, we'll need to introduce the concept of a **variable**. In this book, we'll use the word "variable" to refer to a quantity, such as a count of things, that has a value that we want to keep track of as it changes. In every difference equation, for example, we will always need a variable to keep track of time, which we will name t. There's nothing particularly special about the name t, and in fact variable names can be anything that we want. In this book, we will generally choose variable names that remind us of what the variable stands for or that match names that are commonly used by other ecologists. Aside from Roman letters, variables are often named using Greek letters, such as α, β, and λ, all of which will appear in this book.

By convention, we begin counting time at time zero, which we represent as $t = 0$. From time zero, t will increase by whole numbers to $t = 1$, $t = 2$, and so on. Each of these **time steps** represents the same length of time, but this length can be anything that we choose. In our seed planting problem, it seems natural to set the time steps to 1 minute. In the next few chapters, we'll use time steps of days and years, depending on which is the most natural for the problem.

To track the number of seeds planted over time in our garden, we will need to define one additional variable, representing the number of seeds that have already been planted at time t. By convention, we'll frequently use the variable name N to represent the number of things, such as seeds, in our equations. Because the value of N changes with time in our difference equations, we'll use the name N_t to indicate the count of seeds that we have planted as of time t. For example, we would represent the number of seeds that have been planted by time $t = 0$ as N_0. Because no seeds were already planted when we arrived in the garden, $N_0 = 0$. The value of a variable at time zero is known as its **initial condition**, which is simply the value of the variable before we begin our projections.

Up to this point, we have two variables whose values will change over time: t, representing time measured in minutes since the start of our gardening activity, and N_t, representing the number of seeds that have been planted through the end of time period t. The question above asks us how many seeds we will have planted after 10 minutes, and the answer to this question will be given by our projected value of N_{10}.

Along with these two variables, we will need one additional piece of information to set up our difference equation, which is the number of seeds that are planted per time period. The problem above states that we are able to plant seeds at a rate of 2 seeds per minute, which we will represent with the **parameter** $p = 2$. A parameter is a quantity, such as a seed planting rate, that we will use in our calculations but which remains constant and unchanging over time (the only exception in this book will be in chapter 24, where we will purposefully vary our parameters in a particular way). In other words, unlike the variables t and N_t that change as we make our projections, the value of the parameter p will stay the same. Of course, we may be interested in making projections based on different possible values of our parameter. However, whenever we choose a value for a parameter, that value stays the same throughout the entire projection.

We are now finally ready to set up our difference equation to project the number of seeds that will have been planted as time progresses into the future. Beginning with $N_0 = 0$, we can calculate N_1 by recognizing that exactly p more seeds will be planted in this one time step. We thus know that $N_1 = 2$. If we were to write this logic down as an equation, we would write

$$N_1 = N_0 + p = 0 + 2 = 2$$

Going forward to the next step, it turns out that we can calculate N_2 using the value $N_1 = 2$, which we know from the equation above, in a very similar manner:

$$N_2 = N_1 + p = 2 + 2 = 4$$

Continuing on in this same manner for N_3, N_4, and so on, we'll eventually find that $N_{10} = 20$, matching our mental calculation of the answer from before. Importantly, at each time period, we are able to make a projection for exactly one additional time step into the future based on the known value for the present. In a manner similar to climbing a ladder, we can continue stepping our way into the future, one step at a time, and calculating new values of our variables for as long as we want.

As you can see above, the equation that we need to calculate the value of our variable in the next time step using the value in the current time step always looks the same. We can, in fact, write a more general version of this equation that applies to any time t from 0 to infinity. If we use t to represent the current time step, then $t + 1$ represents the next time step after this one. Looking at the equations above, we can infer that the general equation

$$N_{t+1} = N_t + p$$

would allow us to calculate how many seeds will have been planted by the end of any period $t + 1$ given how many have been planted by the end of t periods.

Of course, for a problem this simple, there's hardly any need to bother writing down an equation at all. The next four chapters will present increasingly more complicated scenarios where the answers to the problems cannot be seen so quickly or easily. In these cases, it will be very helpful to take the time to carefully write down the equations for the system being described, using both variables and parameters as needed, in order to make our projections. The next chapter will show you specifically how to make such a projection in a Google Sheet.

Incidentally, you might be wondering at this point why these equations are called "difference equations." This is because another common way of considering change over time is to focus on calculating the *difference*, or change, in a variable from one time step to another, rather than the actual value of the variable in the next time step. If we look at our equation above for calculating N_{t+1} from N_t, for example, we could rearrange this equation to read

$$N_{t+1} - N_t = p$$

This way of writing the equation makes it clearer that the difference in the number of seeds planted between any two time steps is equal to the constant parameter p.

Readers with a background in calculus or population modeling may be wondering how the difference equations that we're using here relate to the differential equations that are also used for projecting change over time. In essence, we can think of a differential equation as the limit of a difference equation when the time step gets very short. In the version of the difference equation above, the right-hand side of the equation gives the change in the variable N from one time step to the next. In the equivalent differential equation, the right-hand side gives the *slope* of the population projection in continuous time. More information on this relationship, as well as differential equation versions of all of the models presented in part 1, can be found in Otto and Day (2007).

2 **Duckweed on a Pond** Exponential Growth

While on a hike, you pass by a pond and notice a small patch of floating duckweed plants near the bank. The patch is about 10 × 10 cm in area, a small fraction of the approximately 20 × 50 m surface of the pond. You have read that duckweed populations can grow quickly, up to 40% each day under ideal conditions. How quickly could this patch of duckweed expand to cover the entire pond?

As described in the last chapter, difference equations provide a general framework for projecting how populations change over time. The problem in this chapter requires us to make just such a projection. Specifically, we wish to determine how long it would take a 10 × 10 cm patch of duckweed, growing at a rate of 40% each day, to cover the 20 × 50 m surface of the pond.

To begin, we will use the variable N_t to refer to the area of duckweed on the pond, measured in square meters, on day t. In future chapters, we will usually use the variable N_t to refer to a number of individuals, but here we can think of N_t as counting the number of square meters of the pond that are covered by duckweed on day t. The variable N_{t+1} then refers to the number of square meters of duckweed on the pond on the day after t. The fundamental idea of a difference equation is to use the value of N_t to predict the value of N_{t+1}. If we know N_0, which is the value of the variable on the first day, then we can predict the value of N_t for any future time using a day-by-day calculation.

In this example, our duckweed patch starts off at 10 × 10 cm, or 0.01 m², at $t = 0$. We thus start our calculations with

$$N_0 = 0.01$$

Tomorrow, at $t = 1$, this patch will have grown by 40%, according to the information in the problem. We can calculate N_1 as N_0 plus an additional 40% of N_0, or

$$N_1 = N_0 + 0.4 N_0 = 0.01 + 0.4 \times 0.01 = 0.014$$

In other words, after one day has passed, we expect our duckweed patch to grow from 0.01 m² to 0.014 m², an increase of 40%. Starting now with

$N_1 = 0.014$, we can then proceed to calculate N_2, and then after that N_3 and beyond, using the same difference equation approach:

$$N_2 = N_1 + 0.4N_1 = 0.014 + 0.4 \times 0.014 = 0.0196$$
$$N_3 = N_2 + 0.4N_2 = 0.0196 + 0.4 \times 0.0196 = 0.02744$$

By the time the third day has passed, we can see that the duckweed patch will have more than doubled in area from its initial size.

Although we could continue these calculations by hand, there's a basic pattern in the day-to-day growth equations that we can use to solve this problem more efficiently. Looking at how N_1, N_2, and N_3 are calculated, we recognize that we can calculate the area of the duckweed patch on each day on the basis of its area the day before using the equation

$$N_{t+1} = N_t + rN_t$$

Here, N_{t+1} is the area of the duckweed patch on the next day, N_t is the area of the duckweed patch today, and r is the factor by which the patch increases (or decreases) in size from one day to the next. In this problem, $r = 0.40$, which corresponds to a 40% increase in patch area each day. With this basic equation in hand, we can use a spreadsheet to assist us in projecting the area of the duckweed patch as far into the future as we'd like.

> In a new Google Sheet, type the text t in Cell A1 and the text Nt in Cell B1. In Cell A2, enter the value 0, and in Cell B2, enter the value 0.01. These give the starting day number and duckweed area that you observed. In Cell A3, enter the formula =A2+1 and press Enter or Return. Cell A3 will now contain the number 1. Highlight Cell A3, then click and hold the small blue square in the bottom right of the blue box around the cell and drag downward to Row 37. You'll see Column A fill up with the numbers 0 to 35, representing 35 days of duckweed growth.
>
> Next, in Cell E1, enter the text r, and in Cell E2, enter the number 0.4. Cell E2 stores the value of the variable r that we'll use in our calculations. Then, in Cell B3, enter the formula =B2+B2*$E$2. This is the difference equation for updating the area of the duckweed patch from one time step to the next. Highlight Cell B3, click and hold the small square in the bottom right of the cell, and drag down to Row 37. Note that the dollar signs that are part of the reference to

Cell E2 are needed so that when we drag Cell B3 downward, each cell below still refers to the value of *r* in Cell E2.

Projections for the first 10 days, rounded to two decimal places, are shown in table 2.1. At this point, the duckweed patch still covers only about a quarter of a square meter of the pond, which is a fairly small area relative to the area of the whole pond. The complete projection through 35 days, however, shows that the patch will cover the 1,000 m² surface of the pond after 35 days, at which point the patch could have reached an area of 1,302 m² had the pond been that large. Figure 2.1 shows a plot of the duckweed patch area for the first 35 days. Notice that the area of the duckweed patch seems to be very small up until around the 20th day, at which point it increases very rapidly to cover the entire pond.

To create a plot of duckweed area versus time, highlight Column A by clicking on the letter A above Row 1. Then hold Shift and click on the letter B to the right, which will highlight Columns A and B at the same time. Go to Insert → Chart in the Google Sheets menu bar, near the top of the page. A chart will appear on your sheet, and the Chart Editor panel will appear on the right-hand side of the screen. You can use the Chart Editor to change the appearance of the chart.

Table 2.1 Duckweed patch area: projections for 10 days

t	0	1	2	3	4	5	6	7	8	9	10
N_t	0.01	0.01	0.02	0.03	0.04	0.05	0.08	0.11	0.15	0.21	0.29

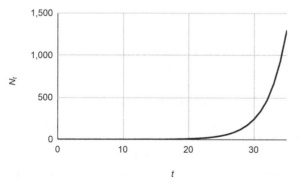

Figure 2.1. Exponential growth of duckweed patch area over time.

The model that we have proposed here for projecting duckweed growth, known as the **exponential growth** model, is one of the simplest possible models of population change over time. Exponential growth assumes that growth between two time steps is always a fixed percentage, regardless of how large the population becomes.

The exponential growth model is so simple, in fact, that we can calculate the projected population size at any point in the future without calculating all of the intermediate population sizes, as we did above. To see how this is done, we can return to the basic equation for exponential population growth, which was

$$N_{t+1} = N_t + rN_t$$

You might notice that this equation could also be rewritten as

$$N_{t+1} = (1 + r)N_t = RN_t$$

The large R variable, known as the **reproductive factor** for the population, is simply equal to $1 + r$. In this problem, we thus have $R = 1.4$.

The advantage to rewriting the equation in this way is that we can see an even simpler pattern when we begin to write our projections for the first few time steps:

$$N_1 = RN_0$$
$$N_2 = RN_1 = R \times RN_0 = R^2 N_0$$
$$N_3 = RN_2 = R \times R \times RN_0 = R^3 N_0$$

This pattern makes it clear that we could calculate the population size at any time t in the future using the equation

$$N_t = R^t N_0$$

Plugging $t = 35$ into the equation above gives the same value of $N_{35} = 1,302$ that we calculated before using the step-by-step calculation.

It's worth noting that the basic assumption of the exponential growth model is that growth will continue at a fixed rate indefinitely. This assumption is often violated in real systems. For starters, once a population reaches the maximum possible size, such as 1,000 m² in this problem, its growth rate r at that point must be zero, since N_t can no longer increase. More subtly, however, in real ecosystems, it often appears that as a population approaches its maximum size, the growth rate decreases, rather than staying the same. We'll explore one very important model of this type in the next chapter.

NEXT STEPS

The exponential growth equation is used widely in many fields of science, and you should have no trouble finding explanations of it in almost any textbook with a modeling section. It's worth knowing that in some ecology papers, particularly those focused on conservation, the reproductive factor R is instead represented by the variable λ, which is the Greek letter lambda. We'll switch to using λ in part 2, as this notation is most common when using matrix models. It's also worth noting that when we are using a difference equation with discrete time steps, exponential growth is also, and perhaps more properly, known as **geometric growth**.

Exponential growth is often assumed to be the basic pattern by which populations will grow in the absence of any resource limitations. The key concept here is to recognize that populations grow by *multiplying* themselves by a constant number in every time period, rather than by *adding* a constant number in every time period. This leads to the rather extreme, potentially explosive power of exponential growth. For a memorable illustration of this principle, look up the "wheat and chessboard problem" or the folktale of "one grain of rice."

3 Throwing Shade I Logistic Growth

> *A few years ago, a large landslide occurred on the side of a nearby mountain, wiping out all vegetation in the area. Researchers have begun studying how one particular plant species is recolonizing the bare ground. In one 2 × 2 m experimental plot, there are now 12 plants of this species, with each individual occupying an area of about 10 × 10 cm. The plants produce many seeds each year, but the researchers believe that only about one viable seed from each adult lands somewhere in the plot each year. These seeds will germinate unless they land under an existing plant, in which case they will be shaded and will not grow. Adult plants have about a 40% chance of dying each year. How long will it take until the plot is completely covered with plants?*

At first, it might appear that this plant population will grow in a similar fashion to the duckweed patch in the last chapter. Let's start building a projection of the future population of this plant using the same approach as we used in the last chapter.

We know that there are currently 12 plants in the population, so we set $N_0 = 12$. In the last chapter, we were told that the duckweed patch would grow by 40% each day. In that example, we didn't stop to think about exactly what causes a population to get bigger or smaller over time. If we think about this for a moment, we can recognize that a population will grow when new individuals are born, and it will shrink when existing individuals die. In this problem, we will consider the population to be **closed**, such that birth and death are the only two ways that the population can change in size. In an **open** population, it's also possible for the population to grow when individuals move into it (immigrate) or to shrink when individuals leave it (emigrate). In the real world, relatively few populations are likely to be truly closed. However, if the number of immigrants or emigrants is small relative to the number of births or deaths, our projections for a closed population will often be very close to those for an open population.

For this closed population, we thus need to think about birth rates and death rates in order to determine how the population will change. According to the problem, each plant produces about one viable seed each year that lands within the plot. If we assume for the moment that these seeds always germinate successfully, we will have one newly born plant in the next year for each adult plant in the population this year. We'll record this number as a per-plant birth rate of $b = 1$. Each adult has a 40% chance of dying each year, leading to a per-plant death rate of $d = 0.4$. We can build a model similar to the one in the last chapter by recognizing that the number of plants in the population next year is equal to the number this year, plus the number born, minus the number that die. This leads to the equation

$$N_{t+1} = N_t + bN_t - dN_t = N_t + (b - d)N_t$$

The quantity bN_t gives the total number of new plants born each year, and the quantity dN_t gives the total number that die each year. This equation is an exponential growth model, just like the one described in the last chapter. The variable r from the last chapter is equal to $b - d$ in the equation above.

Using the same methods as in chapter 2, we find that according to the model above, the plant population would grow quickly. Since each plant occupies 10×10 cm = 100 cm² = 0.01 m², the entire 4 m² plot can hold a maximum of 400 individuals. We can calculate that this population size would be reached at time $t = 8$, or eight years from now.

The plant population described in this chapter, however, does not grow in the same manner as the duckweed population from the last chapter. This problem includes one additional complication regarding the birth of new plants, which is that a seed will not germinate if it lands underneath an existing plant. This self-shading is a type of **intraspecific competition**, or competition between individuals of the same species. In this case, the competition is for space, since no two plants can occupy the same location in the plot.

We can account for this intraspecific competition by modifying the way that we calculate the number of births in the equation that we just proposed. Instead of assuming that the number of births is always the number of seeds per adult times the number of adults, we will allow the number of births to account for self-shading. When there are very few plants in the population, almost every viable seed will land in open space. However, if the plot is nearly completely covered with plants, nearly every seed will land under an existing plant. The actual number of births will then be nearly zero, even though many seeds are produced, since none of the new seeds will germinate.

Let's now figure out exactly how to calculate the number of births. To start, we'll use the variable $s = 1$ to represent the number of viable seeds produced by each adult. The total number of viable seeds produced within the plot each year is equal to sN_t. Some fraction of these seeds, however, land under an existing adult. If the seeds land randomly throughout the plot, then the chance of a seed landing under an adult is simply the number of adults divided by the number of "spots" that could be occupied by adults. Since we have already calculated that a maximum of 400 adults can fit in the plot, we know that there are 400 potential spots for seeds to germinate. We'll use the variable $X = 400$ to represent the number of spots, and thus the chance that a seed lands under an existing adult is N_t/X.

Putting these calculations together, we can see that the actual number of births in this plant population will be the number of seeds produced minus the number of seeds that land under adults. This gives us

$$bN_t = sN_t - sN_t \times \frac{N_t}{X} = sN_t\left(1 - \frac{N_t}{X}\right)$$

Let's say, for example, that the plot currently has 100 plants in it. This means that 100 seeds will be produced this year. However, 25 of those seeds, a number that we calculate from 100 × (100/400), will land under an adult and will not germinate. Therefore, the actual number of births is 75, less than the 100 seeds that were produced.

We can now modify the exponential growth equation to account for this more complex way of calculating the number of births each year:

$$N_{t+1} = N_t + bN_t - dN_t = N_t + sN_t\left(1 - \frac{N_t}{X}\right) - dN_t$$

Interestingly, this difference equation leads to a very different pattern of population growth than our previous exponential growth equation. Figure 3.1 shows that we now project that the population will grow quickly at first, in a manner similar to exponential growth, but that the growth will slow down over time. In fact, after about 14 years, the population settles down to a final size of 240 individuals.

> To project the change in this plant population, create a Google Sheet using the instructions for the exponential growth equation given in the last chapter and then make the following changes. Replace the variable r with the variable s and enter its value of 1 in Cell E2. Add the variable name X to Cell F1 and its value of 400 to Cell F2. Add the variable name d to Cell G1 and its value of 0.4 to Cell G2. In Cell B2, enter 12 for the starting population size. In Cell B3, use the formula

=B2+$E$2*B2*(1-B2/$F$2)-$G$2*B2 to project the size of this plant population into the future. A plot of future population size can also be made following the instructions from the last chapter.

The answer to the problem thus appears to be that the plot will, in fact, never be completely covered with plants! Eventually, the population will level off at 240 individuals, leaving the plot permanently 60% full.

Although the equation above provides the answer to our problem, ecologists often like to write this equation using different variables. To start, we can define a variable $r = s - d$, which represents the population growth rate that we would see if there were no self-shading and the population grew exponentially. Then, we can define another variable K, which represents the long-term steady-state number of individuals that will inhabit the population. It takes a bit of algebra to figure out, but it turns out that $K = (s - d)X/s$. In this problem, for example, we have $K = (1 - 0.4) \times 400/1 = 240$. If we substitute in the new variables r and K for s, d, and X in the equation above, we can rewrite that difference equation as

$$N_{t+1} = N_t + rN_t\left(1 - \frac{N_t}{K}\right)$$

This is the equation for **logistic growth**, which is a widely used model for describing a population whose growth slows over time until leveling off at a steady-state value. In this logistic growth equation, r represents the growth rate of the population when the population is small, or equivalently, the growth rate of the population in the absence of the self-shading process. K represents the **carrying capacity** of the population, which is the long-term number of individuals in the population. Logistic growth produces the classic S-shaped curve shown in figure 3.1,

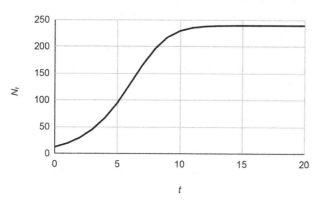

Figure 3.1. Logistic growth of a plant population over time. The population eventually reaches an equilibrium of 240 individuals.

where the population grows almost exponentially at first and eventually levels off at K.

This classic logistic growth model applies to a single species in which individuals compete with each other for some resource, such as space. The next chapter will extend the logistic growth equation to two species that compete with each other.

NEXT STEPS

The logistic growth equation is probably most frequently used to describe changes in the number of individuals in a population, as we've done here. However, this equation crops up in many other contexts in ecology, where it can arise due to similar "self-shading" processes. If we interpret "spots" where adults can grow as entire habitat patches, for example, the logistic growth equation can be used to model changes in the number of habitat patches that are occupied, where s is the rate at which dispersers reach a new patch and d is the rate at which patches become unoccupied. In this context, the logistic growth model is known as the Levins metapopulation model. We can also imagine a disease that can repeatedly infect hosts in a population, where s is the rate at which the disease is spread to a new uninfected individual and d is the rate at which individuals recover. In this interpretation, K is the number of hosts infected at any given time.

It's important to recognize, however, that the logistic equation is not always assumed to result from some type of intraspecific competition or self-shading process. Sometimes this equation is used **phenomenologically**, which means that it's chosen simply because it's a relatively simple equation that produces roughly the kind of curve that's desired. In this context, the logistic equation pops up in logistic regression, which is beyond the scope of this book (but see part 4).

You might take issue with some of the assumptions that lead to the logistic growth equation. For example, the idea that a population will reach a constant number of individuals, never increasing or decreasing from that number, might seem implausible. In the real world, random events are likely to cause the population to move away from its carrying capacity, either upward or downward. (See chapter 24 for ideas on how to simulate this randomness.) The idea, however, is that a population described by the logistic growth model will "try" to return to its carrying capacity over time. You may also have noticed that we assumed that seeds from adults land randomly in a plot, rather than close to the adult plants themselves. Chapter 25 discusses one way of taking spatial position into account.

You might also wonder about an unusual feature of this projection, which is that in most time steps, we calculate that a fractional number of plants will be present in the plot. We should thus consider $N_3 = 29.4975$, for example, to represent approximately 29 or 30 plants in the plot in year 3. If you prefer, you can use the function round() around the formulas in the Google Sheet in order to round the population in each year to the nearest whole number of plants. You'll see that the shape of the projection is essentially the same whether counts in each year are rounded or not. However, it's worth noting that sometimes rounding can affect population projections more significantly, particularly when models are more complex, the values of variables are relatively small, and the parameters of the models are small.

Finally, one last note on our conclusion that the plot will never completely fill up. It turns out that this result is dependent on the order of birth and death in our population. The equations above effectively assume that births occur before deaths within a given year, such that existing plants in year t are able to shade newly generated seeds even if they subsequently die within that same year. In practice, this might correspond to a population that is counted at the start of fall each year. Right after the census, seeds are produced and germinate in the spring, and then mortality of adult plants occurs in the following summer. The model thus states that the plot is never full at the start of fall. This is logical, since plants have died since the previous spring and have not yet had the opportunity to be replaced by newly germinated seeds. This model setup is known as a **pre-breeding census**. An alternative **post-breeding census** model would use a census taken after seeds germinate. You might try developing a different model for this population in which seeds are produced in fall and germinate in spring, adult mortality occurs over the winter, and a census is taken at the end of spring.

4 Throwing Shade II Lotka-Volterra Competition

> *This year's survey of the experimental plot from chapter 3 was just completed, and one of the students working on the project noticed that a threatened plant species, previously overlooked, has a small population of 5 individuals in the plot. The researchers were surprised but excited to find this species in the plot. The threatened plant is about one-quarter the size of the more noticeable colonizing plant from the last chapter. Adults of the threatened plant species produce one viable seed per year and have an adult death rate of 60% each year. The seeds of both plants are shade intolerant, and neither type of seed will germinate if it lands underneath an adult of either species. Is it possible for both the threatened species and the colonizing species to coexist in this plot in the long term?*

This problem poses a classic question about **interspecific competition**, which is competition that occurs between two species. In this community, the two plant species compete with individuals of their own species and individuals of the other species for space. One question to ask in any such situation is whether it is possible for the two species to coexist in the long term, or whether one species will inevitably outcompete the other.

It will help to begin this problem by building population growth models for each of the two species if they were to occur separately. Let's consider first the colonizing plant from the last chapter. We already determined that, when this colonizing plant is the only species present, its population will grow according to the equation

$$N_{t+1} = N_t + s_N N_t \left(1 - \frac{N_t}{X_N}\right) - d_N N_t$$

where $s_N = 1$ is the number of seeds produced per year by an adult plant, $X_N = 400$ is the maximum number of adult plants that can fit in the plot, and $d_N = 0.4$ is the death rate for adult plants. We added the subscript N to these parameters to indicate that these variables refer to the colonizing plant, whose population we will track using the variable N_t.

We can then follow the exact same logic to determine how the population of the threatened plant would grow if it were the only species present. Because seeds of this threatened species have the same response to self-shading as those of the colonizing species, we find

$$M_{t+1} = M_t + s_M M_t \left(1 - \frac{M_t}{X_M}\right) - d_M M_t$$

Here, we've used M_t to represent the number of individuals of the threatened plant species, and the M subscript indicates parameters that refer to the threatened plant. From the problem, we know that $s_M = 1$ and $d_M = 0.6$. Since the threatened plant is one-quarter the size of the colonizing plant, the maximum number of threatened plants that could fit in the plot is four times larger than the maximum number of colonizing plants. Thus, $X_M = 1{,}600$.

The question posed in this problem, however, is about what happens when the two plants occur together in the same plot. When the two species co-occur, the problem states that adults of each plant species can shade seeds of both their own species and the other species. We need to account for this interspecific competition to predict how the populations of the two species will change together.

To do this, we will first think more carefully about the meaning of the fraction N_t/X_N in the first equation. This fraction represents the proportion of the habitat that is currently covered by adult colonizing plants. One minus this fraction thus represents the fraction of the habitat that is *not* covered by adult colonizing plants, which is the fraction of the plot in which a seed is able to germinate. Multiplying this fraction by $s_N N_t$, which is the total number of seeds of the colonizing species produced in year t, gives the total number of seeds of the colonizing species that will germinate.

The next step is to modify the population growth equation for the colonizing species to account for the presence of adults of the threatened species. With the threatened species present, the fraction of the plot that is not covered by adult plants, and hence available for seeds of the colonizing plant to germinate, is now one minus the fraction covered by adults of the colonizing species, minus the fraction covered by adults of the threatened species. We can thus modify the difference equation for the colonizing species to read

$$N_{t+1} = N_t + s_N N_t \left(1 - \frac{N_t}{X_N} - \frac{M_t}{X_M}\right) - d_N N_t$$

Throwing Shade II: Lotka-Volterra Competition

By identical logic, we need to modify the difference equation for the threatened species to read

$$M_{t+1} = M_t + s_M M_t \left(1 - \frac{M_t}{X_M} - \frac{N_t}{X_N}\right) - d_M M_t$$

Together, these two equations give us enough information to project the populations of both plant species into the future. Notice that we now need to calculate the values of two variables in each time step, N_{t+1} and M_{t+1}, based on the values of N_t and M_t from the previous time step.

> Begin with a Google Sheet set up identically to the one from chapter 3. Replace the variable names s, X, and d in Cells E1 to G1 with sN, XN, and dN. Then add the variable names sM, XM, and dM in Cells H1 to J1, and the values 1, 1,600, and 0.6 in Cells H2 to J2. Add the variable name Mt to Cell C1 so that we can track the population of the threatened plant. Enter the initial size of the threatened plant population, which is 5, in Cell C2.
>
> Next, modify the formula in Cell B3 to read =B2+$E$2*B2*(1-B2/$F$2-C2/$I$2)-$G$2*B2, which will calculate the number of colonizing plants while accounting for the fraction of the plot covered by the threatened plant. Similarly, enter the formula =C2+$H$2*C2*(1-C2/$I$2-B2/$F$2)-$J$2*C2 in Cell C3 to calculate the number of threatened plants. Highlight Cells B3 and C3 and drag the bottom right corner of C3 down to Row 37.

When we make this projection, we find the result shown in figure 4.1. Both populations start off increasing as the plot fills up with plants. However, around year 7, the population of the threatened plant peaks, and then declines thereafter. The threatened plant is eventually eliminated from the plot around year 27, at which time the colonizing plant reaches a steady-state population of 240 individuals, the same population size that we found in the last chapter. These results suggest that in fact, these two species are not able to coexist, as the colonizing plant will eventually outcompete the threatened plant. Interestingly, the long-term fate of the threatened plant may not be clear to an observer watching the plot for the first few years, while both populations are increasing. Our model, however, allows us to anticipate the eventual **competitive exclusion** of the threatened plant from this plot.

It turns out that with some additional rearrangement of the equations above, we can find a general rule that defines the conditions under

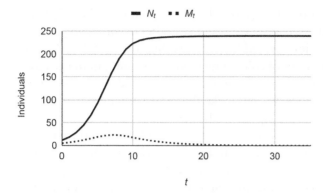

Figure 4.1. Projected populations of a colonizing plant species (N_t) and a threatened plant species (M_t). Although both populations increase initially, the threatened species is eventually excluded from the plot.

which two species are able to coexist. To begin, we'll define a new variable α_{NM}, which is defined as the number of individuals of the threatened species that are equivalent, in competitive terms, to one individual of the colonizing species. Since the two species compete for space, and the threatened plant is one-quarter the size of the colonizing plant, $\alpha_{NM} = 0.25$. In other words, it takes four threatened plants to have the same effect as one colonizing plant on changes in the colonizing plant population from year to year. Similarly, we can define $\alpha_{MN} = 4$, which says that one colonizing plant has the same effect on the population of threatened plants as four threatened plants.

Mathematically speaking, the way to calculate these α variables is to take a ratio of X_N and X_M. For example, the number of individuals of the threatened species that are equivalent to one individual of the colonizing species can be calculated as $\alpha_{NM} = X_N/X_M = 0.25$. Similarly, we can see that $\alpha_{MN} = X_M/X_N = 4$. If we substitute these two variables into the difference equations for the population of each species, we can then write these equations as

$$N_{t+1} = N_t + s_N N_t \left(1 - \frac{N_t + \alpha_{NM} M_t}{X_N}\right) - d_N N_t$$

and

$$M_{t+1} = M_t + s_M M_t \left(1 - \frac{M_t + \alpha_{MN} N_t}{X_M}\right) - d_M M_t$$

It may take a bit of staring at the equations above to figure out how they work, but the basic idea is the same as in the versions that we saw earlier. For example, in this new equation for N_{t+1}, the amount of the plot covered by plants is now $N_t + \alpha_{NM}M_t$, which is the total number of plants of both species expressed as an equivalent number of colonizing plants. The fraction of the plot covered by plants is then this number divided by the maximum number of colonizing plants that could fit in the plot.

Finally, to reach the form of these difference equations most commonly used by ecologists, we can make the same variable substitutions that we made in the last chapter. We'll once again define the variables $r_N = s_N - d_N$ and $r_M = s_M - d_M$ as the maximum growth rates of the colonizing and threatened species, respectively, if there were no shading effects on the seeds of either species. We can then define $K_N = (s_N - d_N)X_N/s_N$ and $K_M = (s_M - d_M)X_M/s_M$ as the carrying capacity for each species, or equivalently, as the number of individuals of each species that would be found in the plot in the long term if that species were the only one in the plot.

Substituting in these new variables and doing some algebra leads to our final set of equations:

$$N_{t+1} = N_t + r_N N_t \left(1 - \frac{N_t + \alpha_{NM} M_t}{K_N}\right)$$

$$M_{t+1} = M_t + r_M M_t \left(1 - \frac{M_t + \alpha_{MN} N_t}{K_M}\right)$$

This particular way of expressing our two difference equations forms the famous **Lotka-Volterra competition** model for two species. This chapter hopefully makes it clear that this model is a logical and direct extension of the logistic growth model for a single species, where now the change in the population of each species is modified to account for an "equivalent" number of competing individuals of other species. The **competition coefficients**, given by the α parameters, describe this equivalence.

It turns out that these equations provide a very simple and general rule for the coexistence of two species. Although it's beyond the scope of this chapter to derive the basis for this rule, the rule states that two species can coexist only when both of the α variables are less than 1. In other words, for species to coexist in this model, the individuals of each species must compete more strongly with individuals of their own species than with individuals of the other species. Because this criterion is not met in this problem, the two species here are not able to coexist in the long term. You can verify that this competitive exclusion occurs regardless of how many individuals of the two species are found initially in the plot.

NEXT STEPS

The Lotka-Volterra model is a very widely used framework for modeling interactions between species, including community-level food webs and other interaction networks. Interestingly, by changing the signs on the α variables, these equations can be used to model not only competitive relationships, but also mutualistic, parasitic, or commensal ones. The textbook by Otto and Day (2007) contains a comprehensive overview of how to construct and analyze models like these, as well as their differential equation equivalents in continuous time.

As the Lotka-Volterra model is essentially a multi-species extension of logistic growth, many of the same caveats and extensions that apply to the logistic growth model described in chapter 3 also apply here. In particular, note that Lotka-Volterra models are often used phenomenologically, as a description of a general pattern of interactions, rather than being derived from the particular details of interspecific competition for a particular resource.

5 Rabies Removal SIR Models

> *A fox population of about 1,000 individuals lives in the landscape surrounding the experimental plot described in the last two chapters. A local rancher recently spotted a fox on her property behaving in a way that suggested a rabies infection. Rabies is an inevitably fatal disease, and infected foxes have about a 50% chance of death each day after infection. Rabies spreads by contact between foxes, and an infected individual might have about a 1 in 1,000 chance each day of encountering an uninfected individual and transmitting the disease. If only this one fox is currently infected, could the disease eventually spread through the population? If so, how many foxes might eventually die of rabies?*

This problem asks whether rabies is likely to become an **epidemic**, eventually spreading through the fox population and potentially killing a large number of currently uninfected individuals. If an epidemic does occur, the problem asks specifically how many foxes might die of the disease due to the initial presence of one rabid fox.

The problem makes it clear that both infection and death following infection occur relatively rapidly. We will thus model the potential outbreak of rabies in this population using time steps of days. Since we will find that the course of this outbreak proceeds quickly, we will assume for simplicity that there are no births or deaths in the population during this period due to factors other than the disease. These are common simplifications for modeling potential epidemics that occur over the course of a season, rather than over multiple years.

On any given day, the fox population can be divided into three groups, or classes, which we will track using three different variables. The classes are foxes that have not been infected and are susceptible to the disease, S; foxes that are currently infected, I; and foxes that are removed from the population by death, R. Reflecting these variable names, disease models of this nature are often referred to as **SIR models**. It's worth knowing that the variable R usually stands for individuals that have recovered and are immune to future infection, rather than individuals removed from the population, as in this problem. To determine how many foxes will eventually die of the disease, we need to determine how large R will eventually become. If it were to reach 1,000, the entire population would be wiped out.

We can begin by writing difference equations for each of the three classes. The number of foxes in the susceptible state is initially $S_0 = 999$, since one fox is currently infected. This value will decrease over time as susceptible foxes become infected. We also begin with $I_0 = 1$, since we are assuming that the one infected fox is the only infected individual, and $R_0 = 0$, since the disease has just arrived in the population and no individuals have died of the disease yet.

We also know from the problem that each infected fox has a 0.1% chance of infecting a susceptible individual each day. We'll use the variable $\beta = 0.001$ to reflect this infection rate. We thus expect each individual infected fox to infect βS_t susceptible foxes on day t. The total number of new infections per day is thus $\beta S_t I_t$. The total number of susceptible individuals on day $t + 1$ must thus be the number of susceptible individuals on the previous day t, minus the number of newly infected individuals that move from the susceptible to the infected class. The difference equation for the number of susceptible individuals is thus

$$S_{t+1} = S_t - \beta S_t I_t$$

Since every newly infected individual moves into the infected class, the value of I_{t+1} must then grow each day by this same number, $\beta S_t I_t$. The number of foxes in the infected class, however, also decreases each day due to death. The number of infected foxes who die each day is equal to dI_t, where $d = 0.5$, as described in the problem. We can then write a difference equation for the change in the infected fox class as

$$I_{t+1} = I_t + \beta S_t I_t - dI_t$$

The last class in this population is the "ghost" class of foxes that have died due to the disease and are no longer present in the living population. The count of removed foxes grows each day by the number of infected foxes that die in the previous day. The difference equation for this class is thus

$$R_{t+1} = R_t + dI_t$$

This gives the final combined model for all three classes:

$$S_{t+1} = S_t - \beta S_t I_t$$
$$I_{t+1} = I_t + \beta S_t I_t - dI_t$$
$$R_{t+1} = R_t + dI_t$$

These three equations have an interesting property that can be seen clearly if we add them together. Adding the left-hand and right-hand sides of these equations all together, we find that all of the terms that add or subtract individuals from the classes cancel out, leaving us with

$$S_{t+1} + I_{t+1} + R_{t+1} = S_t + I_t + R_t$$

This equation reflects the fact that the total number of foxes in this population does not change over time. Notice, however, that in this calculation we continue to include the "ghost" foxes removed from the population in the total count of 1,000 individuals.

> In a new Google Sheet, enter the variable names t, St, It, and Rt in Cells A1 through A4. In Cells A2 through A62, enter the numbers from 0 to 60, which count the days of the potential epidemic. In Cells B2 through D2, enter the values 999, 1, and 0, which give the initial number of individuals in the susceptible, infected, and removed classes. In Cell F1, enter the variable name Beta, and enter its value of 0.001 in Cell F2. In Cell G1, enter the variable name d, and enter its value of 0.5 in Cell G2.
>
> Next, in Cells B3 through D3, we'll enter the difference equations to update our calculations of the count of individuals in each of the three classes from day to day. In Cell B3, enter the formula =B2-$F$2*B2*C2. In Cell C3, enter the formula =C2+$F$2*B2*C2-$G$2*C2. Finally, in Cell D3, enter the formula =D2+$G$2*C2. Highlight these three cells, and drag down from the bottom right corner of Cell D3 to Row 62 to complete the projection.

The three difference equations above can be used to project how this disease will eventually spread through the fox population. With initial class populations of $S_0 = 999$, $I_0 = 1$, and $R_0 = 0$, figure 5.1 shows that rabies does become an epidemic in this population. There is a peak of 174 infected individuals at day 16, after which the number of infections begins to decrease. After about a month, the number of individuals in the R class plateaus near 829. The rabies epidemic thus leads to the deaths of 829 foxes, leaving 171 survivors in the population once the epidemic has passed.

Intuitively, we might expect that it would be hard to know the exact values of the parameters β and d for this population, or for any population. We might thus be interested in seeing how the values of β and d affect

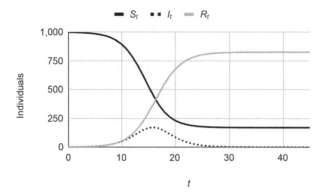

Figure 5.1. Projected sizes of the susceptible (S_t), infected (I_t), and removed (R_t) classes through the course of the epidemic.

the model predictions. If β were in fact smaller, meaning that rabies was transmitted more slowly between foxes than we originally thought, the disease would spread more slowly, and fewer foxes would die of the disease. For example, if we use β = 0.0001, we predict that the disease will not spread at all, and that only the originally infected individual will die. Conversely, if we use the original value of β = 0.001 with a decreased death rate of d = 0.1, we predict that each infected fox will infect a much larger number of susceptible foxes before dying. In this scenario, the entire population is either infected or dead by day 17. Changing the two parameters to other values will give different projections for the disease outbreak.

Beyond making specific projections, we can also draw a very general conclusion from these equations themselves. The difference equation for the change in the number of infected individuals from the initial number to day 1 is

$$I_1 = I_0 + \beta S_0 I_0 - d I_0$$

For the disease to begin to spread through the population, the number of infected individuals must grow over time. We can see from the equation above that the number of infected foxes will increase only when $\beta S_0 I_0 > d I_0$. In SIR models like this one, this criterion is known as the **epidemic threshold**. This criterion suggests several actions that can be used to limit the spread of a disease when intervention is possible, as in human populations. For example, β can be reduced by limiting contact between infected and susceptible individuals, and S_0 can be reduced by vaccination. When infected individuals eventually recover and achieve immunity from a disease, d represents the recovery rate, which can sometimes be increased by treating the disease effectively.

NEXT STEPS

If you tried exploring different values of β and d in this model, you may have noticed something unusual happening if you made the parameter β too high. If you do this, you'll find that the model will "crash," projecting negative numbers and then enormous numbers of foxes in different classes. In fact, if you return to chapter 3, you can induce the same behavior in the logistic growth model by making the parameter r large enough.

This behavior highlights a restriction on the use of difference equations for models like this one, which is that the rate parameters, such as β and d, must be relatively small for the model to create smooth projections. If, in fact, the number of foxes infected per day by a single infected individual was much higher, we could avoid the problem of the model "crashing" by making the time steps smaller—for example, hours instead of days. If you imagine continuing down this path of making the time steps smaller and smaller, you'll eventually arrive at a differential equation, which is equivalent to a difference equation in the limit in which the time step becomes infinitesimally small. Differential equations are analyzed using the tools of calculus, which are not part of this book.

This problem and the answer to it are both based on a paper by Anderson et al. (1981), who develop a similar but more detailed model of rabies in fox populations of Europe.

Part II Understanding Uncertainty

6 Introducing Probability

In part 1 of this book, all of the values of our variables, as well as our predictions of their values in the future, were known exactly. For example, when we stated that the size of a patch of duckweed was 0.01 m^2, or that the population of a plant species would eventually reach 240 individuals, we assumed that these numbers were all known without any error or uncertainty. In the real world, however, the information that we have is messy, and we may not be entirely confident that we are able to know or predict the exact values of the variables that we are interested in.

The idea of understanding and quantifying uncertainty in the values of our variables lies at the heart of probability, which is the topic of this part of the book. To get started with probability, we will introduce a new type of variable, which is known as a **random variable**. Unlike the variables in part 1 of this book, a random variable is a variable whose value is described not by a single value, but by a set of possible values, each with an associated chance of being observed.

Let's make this idea more concrete with an example. Imagine that we have a standard, fair, six-sided die that we roll six hundred times. If the die is fair, we understand intuitively that we should expect to see each of the faces, with one through six dots, come up about one hundred times each. The number of dots on top of the die is a random variable, which we can call D. We can then define the **probability** of D taking any value as the fraction of times that it will take this value if the die is rolled many, many times, approaching an infinite number of times.

If we were able to actually roll the die an infinite number of times, we would expect that each of the six numbers would appear on top of the die in 1/6 of the rolls. We thus know that the probability of observing each number is 1/6. We write down each of these probabilities as

$$\Pr(D = 1) = 1/6$$
$$\Pr(D = 2) = 1/6$$
$$\Pr(D = 3) = 1/6$$
$$\Pr(D = 4) = 1/6$$
$$\Pr(D = 5) = 1/6$$
$$\Pr(D = 6) = 1/6$$

Together, these six probabilities form a **probability distribution**, which is the list of all possible values that a random variable can take along with the probabilities of observing those values. In this case, the possible values are the numbers one through six, and the probability of observing any of these values is 1/6. The set of possible values that a random variable can take is known as the **support** of that distribution. This probability distribution can be further described as a **discrete probability distribution** because the random variable can take only a set of whole-number values, each associated with a number of dots. We'll work entirely with discrete probability distributions in this book.

If we were to add up the six probabilities listed above, we would find that they sum to 1. This has to be true by definition, since each probability represents the proportion of times that a random variable takes a certain value. If we were to sum all of these proportions across all possible values for the random variable, we would always get a total of 1. This requirement is one of the most basic rules of probability, which is that the sum of probabilities across all possible values of a random variable must equal 1. This rule means that if we sum the values of a probability distribution across the entire support of the distribution, we will always find that this sum equals 1. We'll use this rule repeatedly in the next few chapters, particularly in chapter 8.

Up to this point, we have only talked about the probabilities associated with a single random variable, D. Just as we were able to create models with multiple variables in part 1, we can consider situations in which we have multiple random variables. We might, for example, have a second die that we roll at the same time as the first, or have two people roll a single die one after the other. In either case, we could use the random variable E to represent the number that appears on the second die roll.

A very important concept in multi-variable probability is the idea of **independence** between variables. Two variables are independent if the probability of observing a particular value of one random variable does not depend on the value of the other random variable. In the example above, it is very reasonable to assume that the values of the two random variables D and E are independent, since the number that comes up on top of the first die shouldn't affect the probability of any particular number coming up on top of the second die. We'll tackle some multi-variable problems and discuss the idea of independence further in chapters 8 and 10.

Before moving on, it's worth pausing for a moment to point out that the definition of probability given above is a **frequentist** definition. As the name suggests, frequentist probability is defined as the *frequency* with which a random variable takes a particular value. There is an entirely different definition of probability that is also in widespread use, which is **Bayesian** probability. Bayesian probability is defined as an observer's belief in the chance that a random variable will take a certain value. The differences between frequentist and Bayesian probability, which lead to fairly different approaches to statistics, are beyond the scope of this book. You can learn more about these differences in many books, including in Bolker (2008). We'll focus exclusively on frequentist probabilities in part 2 of this book, and then exclusively on frequentist statistics in part 4.

7 A Bird in the Cam I Single-Variable Probability

> *A local nature museum would like to install a camera to stream real-time video of an eagle nest in the landscape surrounding the museum. The museum is hoping to install the camera at a nest with at least two eggs in it. A published study on nesting success of this eagle species in a similar habitat reported finding 4 nests with one egg, 14 nests with two eggs, and 2 nests with three eggs. What is the probability that the camera team will find two or more eggs in the first nest that it examines? What is the probability that the team will look at three nests in a row and find only one egg in each? What is the average number of eggs per nest of this eagle species?*

This problem will give us the chance to explore the basic principles of frequentist probability. In the last chapter, we defined probability as the number of observations in which a random variable takes a particular value divided by the total number of observations. That is, the probability of a variable taking some value is equal to the proportion of all observations in which that value is observed as the number of observations becomes very large, approaching infinity. In this problem, the random variable will be the number of eggs found in an eagle nest, N. We will presume that the camera team chooses nests at random and climbs up to observe the number of eggs in each nest.

The first part of the problem asks for the probability that a single eagle nest contains two eggs. We write the probability that the nest contains two eggs as $\Pr(N = 2)$. To determine this probability, we will assume that the probabilities of finding particular numbers of eggs per nest in our landscape near the nature museum are the same as the proportions reported in the published study. We should keep in mind that this is something of an approximation, because the published study did not observe an infinite number of nests. For example, if we were to look at many more nests, it might be possible to find a nest with more than three eggs, even though none were found in that particular study. In chapter 17, we'll discuss how we can use statistical methods to help us think through these kinds of possibilities. For now, however, we'll assume that the frequencies reported in the study are equal to the probabilities of observing different numbers of eggs in a nest in our landscape.

With this assumption, we can calculate the probability that two eggs will be observed in a nest as $\Pr(N = 2) = 14/20 = 0.7$, or 70%. The probabilities of observing one or three eggs can be calculated similarly, giving the probabilities of finding any number of eggs from one through three as

$$\Pr(N = 1) = 4/20 = 0.2$$
$$\Pr(N = 2) = 14/20 = 0.7$$
$$\Pr(N = 3) = 2/20 = 0.1$$

Notice that these three probabilities together add up to 1, as required by the basic rule of probability described in the last chapter.

Using these probabilities, we can answer the first part of the problem. The probability that a nest contains two or more eggs can be calculated by adding together the probability that it contains two eggs and the probability that it contains three eggs. Notice that we don't need to account for the possibility of the nest containing more than three eggs, since numbers greater than three are not part of the support of our probability distribution. This sum is equal to 0.8, which translates to an 80% chance of any randomly chosen nest having two or more eggs. This example illustrates a very general principle of probabilities, which is that we can *add* the probabilities of observing particular values of a random variable together to calculate the probability of observing any one of those values.

The second part of the problem asks for the probability that the team visits three nests in a row and is disappointed to find that all three nests contain only one egg. We have already calculated the probability of observing a single nest with one egg as $\Pr(N = 1) = 0.2$. To extend this probability of observing one egg once to observing one egg three times in a row, we'll assume that the three observations are independent of one another. As defined in the last chapter, independence means that the probability of observing some number of eggs in one nest is not affected by the number of eggs that we observe in any other nest.

The assumption of independence appears very frequently in textbook examples of probability, like rolling dice or flipping coins. It's tempting to think that independence is a common property of the real world. Unfortunately, it's easy to think of reasons why our observations in ecological settings might not be independent. For example, if eagle pairs compete for food, a single nest that contains many eggs may lower the probability that another nest contains many eggs. Alternatively, there may be year-to-year variation in environmental conditions that lead all nests to have more or fewer eggs simultaneously. As is commonly the case in the real world, however, we don't have any particular information about the possible non-independence between our observations in this problem. As is also commonly the case, we'll thus make the tentative assumption of independence between the numbers of eggs observed in each nest.

With the assumption of independence, we can now answer the second question, which is the probability of observing only one egg in three nests in a row. When observations are independent, we can *multiply* their probabilities to find the probability of all three observations occurring simultaneously. Imagine, for example, that the probability of flipping a coin and observing "heads" is 0.5. If we flip the same coin twice in a row, the probability of observing two heads is 0.5 × 0.5 = 0.25, since the two flips are independent. Similarly, if we consider $\Pr(N = 1) = 0.2$ to be equivalent to flipping "heads," then the probability of observing three single-egg nests in a row is 0.2 × 0.2 × 0.2 = 0.008, or just under one-tenth of one percent. It thus seems fairly likely that the camera crew will find a nest with more than one egg if they're able to examine three nests.

Finally, the third part of the problem asks about the average number of eggs per nest in this eagle population. You may recall that the average of any set of numbers is simply the numbers added together divided by the count of the numbers you've added. Let's start by using the 20 observations in the published study to calculate the average number of eggs per nest. From the problem, we know that the study found 4 nests with one egg, 14 nests with two eggs, and 2 nests with three eggs. We can calculate the average of these 20 numbers by adding 4 ones plus 14 twos plus 2 threes. Rather than writing out all 20 of those numbers with plus signs between them, we can write the same calculation in shorthand as

$$\frac{4 \times 1 + 14 \times 2 + 2 \times 3}{20} = \frac{38}{20} = 1.9$$

We thus expect that nests of this species contain an average of 1.9 eggs per nest.

It's worth remembering that it can be tempting to interpret an average number of 1.9 eggs as indicating that we "should" or "will probably" find two eggs if we visit a nest. In this example, this is indeed the case, as we have a 70% chance of finding two eggs in a nest. However, imagine instead that the study had found 9 nests with one egg and 11 nests with three eggs. The average number of eggs per nest in this alternative world would have been

$$\frac{11 \times 1 + 0 \times 2 + 9 \times 3}{20} = \frac{38}{20} = 1.9$$

The average number of eggs per nest is exactly same as in the example from our problem. However, even though the average number of eggs per nest in this world is very close to two, we actually have zero probability of observing a nest with two eggs! It is not always the case that most of the observations will lie close to the average. To understand the probability of observing any particular value of a random variable, the full probability distribution is needed, not just the average.

It's relatively common to calculate average values of a random variable, like the number of eggs per nest, from a set of observations of that variable, like the numbers of eggs in 20 observed nests. However, it's also useful to know that the average of a random variable, also known as its **mean** or **expected value**, can be calculated directly from a probability distribution. Looking at the equation that we used to calculate the average number of eggs per nest,

$$\frac{4 \times 1 + 14 \times 2 + 2 \times 3}{20} = \frac{38}{20} = 1.9$$

we can see that we could rewrite it as

$$1 \times \frac{4}{20} + 2 \times \frac{14}{20} + 3 \times \frac{2}{20} = 1.9$$

In this form of the equation, the first number in each term on the left-hand side is a particular number of eggs, and that number is multiplied by the number of nests with that number of eggs divided by the total number of nests. However, this second part of each term is simply the probability of observing a nest with each associated number of eggs. Each term could be written as $i \times \Pr(N = i)$, where i is a new variable representing the number of eggs observed. The entire expression for the mean number of eggs in a nest is thus

$$1 \times \Pr(N=1) + 2 \times \Pr(N=2) + 3 \times \Pr(N=3) = \sum_{i=1}^{3} i \Pr(N=i)$$

The large Greek letter sigma, written as Σ, represents a **summation**, which is just a fancy way of stating that a whole bunch of similar terms should be added together. In this case, the summation says that we should calculate the value of $i \times \Pr(N = i)$ for all values of i ranging from 1 to 3, and then add these together. This is exactly what we wrote down on the left-hand side of the equation above.

NEXT STEPS

As described earlier in this chapter, the assumption of independence is widely used in calculating the probabilities of multiple observations all occurring together. While this assumption is useful, and may often be approximately if not exactly true, there are many ways in which it can be violated in the real world. Perhaps the most important for ecologists is that observations that occur close together in space or time are often more similar to one another than are observations that occur farther apart. Or, in terms of probability, observations that are closer together in space or time are less likely to be independent of one another. For example, while the numbers of eggs in two nests separated by 1,000 km might be independent of each other, what about the numbers of eggs in two nests separated by 1 km? You may be familiar with a variation of this rule, which says that an observation that is part of a group of observations is likely to be similar to other observations in that group. This is the basis of statistical techniques based on blocking that are used to avoid pseudoreplication, which occurs when non-independent data points are incorrectly assumed to be independent.

This chapter also discusses the common assumption that most observed values of a random variable are likely to be close to the average value. This intuition will be true when the probability distribution has a peak, or mode, at this average value with nearly all of the total probability represented by values close to this average. This intuition is probably based on the assumption of a normal distribution, which has the classic "bell curve" shape and is often observed or assumed to describe real-world data. As described earlier, however, flat or U-shaped distributions can have the same mean as a normal distribution while suggesting very different patterns of observations.

The probabilities of eagle nesting success in this problem are based on studies cited in McGahan (1968) for golden eagle nests.

8 A Bird in the Cam II Two-Variable Probability

> *The day before the camera crew begins searching for a nest, you reread the published study on eagle nesting success mentioned in the last chapter and learn some additional information. It turns out that there is a relationship between the number of eggs in a nest and the type of tree in which the nest is found. Out of the 20 nests surveyed in that study, 15 were found in coniferous trees and 5 were found in broad-leaved trees. Out of the 5 nests in broad-leaved trees, 2 contained only one egg. Which kind of tree should the camera crew climb to have the highest probability of finding a nest with two or more eggs?*

At first, this problem might seem like a jumble of numbers, and it might not be clear how to begin. In the last chapter, we had only one random variable, the number of eggs per nest, N. This problem adds a second random variable, which we will call T, describing the type of tree in which a nest is found. The variable T can take two possible values, coniferous or broad-leaved. Solving this problem requires us to calculate the probability that the two random variables N and T take some particular values simultaneously. Rather than a single-variable problem like the last chapter's, we are now solving a two-variable problem in which we are interested in the potential for **interactions** between the two variables.

To make sense of these two random variables and their interactions, we'll organize all of the information that we have about the types of trees and numbers of eggs in a **contingency table**. A contingency table is simply a table in which the possible values of one random variable are shown in columns and the possible values of the other random variable are shown in rows. In this problem, we can set up a contingency table that looks like this:

	$N = 1$	$N \geq 2$
T = Conif		
T = Broad		

In this table, we've reduced the possible values of the variable N from three possibilities, which were one, two, or three eggs, to only two possibilities, one egg or two or more eggs, since we are not interested in distinguishing between nests that contain two and three eggs.

After setting up the contingency table, the next step is to start filling in all the information that we have. Rather than starting with the numbers in the cells of the table, which will show us the numbers of nests having a particular combination of number of eggs and tree type, we'll actually begin by writing numbers below and to the right of the table. These numbers on the outside of the table will be the total numbers of nests with each value of each random variable, without considering the value of the other. We know from the last chapter, for example, that 4 of the 20 nests contained one egg, and that the other 16 nests contained two or three eggs. We can write these numbers below the table in the appropriate columns. We also know from the problem in this chapter that 15 of the surveyed nests were in coniferous trees and that 5 were in broad-leaved trees, so we can write these numbers to the right of the table in the appropriate rows.

	$N = 1$	$N \geq 2$	
T = Conif			15
T = Broad			5
	4	16	

So far, we have written down information about the numbers of nests having particular numbers of eggs and the numbers of nests found in particular types of trees. Next, we need to enter information about the numbers of nests having particular values of N and T simultaneously. From the problem in this chapter, we know that there were exactly 2 broad-leaved trees containing nests with one egg. This allows us to fill in one of the cells inside the contingency table:

	$N = 1$	$N \geq 2$	
T = Conif			15
T = Broad	2		5
	4	16	

Interestingly, knowing this one number in the interior of the table allows us to fill in the other three empty cells. We know, for example, that there were exactly 4 nests in the study that contained one egg. We also know that 2 of these nests were found in broad-leaved trees. That means the other 2 nests with one egg were found in coniferous trees. We can thus add another number to the table:

	$N = 1$	$N \geq 2$	
T = Conif	2		15
T = Broad	2		5
	4	16	

We can continue on to fill in the rest of the contingency table. Since there were 5 nests in broad-leaved trees and 2 of these nests had one egg, the other 3 nests in broad-leaved trees must have had two or more eggs. Since there were 15 nests in coniferous trees and we know that 2 of them had one egg, the other 13 nests in coniferous trees must have had two or more eggs. We can thus complete the contingency table as follows:

	$N = 1$	$N \geq 2$	
T = Conif	2	13	15
T = Broad	2	3	5
	4	16	

We now have a complete count of the numbers of nests with each possible combination of tree type and number of eggs. Since exactly 20 nests were surveyed in the study, we can convert each of these counts into a probability by dividing it by 20.

	$N = 1$	$N \geq 2$	
T = Conif	2/20 = 0.10	13/20 = 0.65	15/20 = 0.75
T = Broad	2/20 = 0.10	3/20 = 0.15	5/20 = 0.25
	4/20 = 0.20	16/20 = 0.80	

Once again, as in the last chapter, we are using the frequencies observed in the published study to approximate the probabilities of observing certain combinations of values of the random variables N and T in our landscape.

Up to this point, all that we've done is take the information from the problems in this chapter and chapter 7 and write it in the form of a table.

Now, using this table, we can turn to determining which type of tree, coniferous or broad-leaved, the camera crew should climb. To answer that question, we'll need to distinguish three different types of probabilities that we can find in our contingency table. These three types of probabilities are important to any problem involving two or more random variables.

The first type of probability that we can see in this table is **marginal probability**. Marginal probabilities, as the name suggests, are the ones written in the margins of the table, just outside of the four cells. Marginal probability is the probability that one random variable takes a particular value, ignoring any other random variables. For example, the marginal probability of a randomly chosen nest from the study having only one egg is 0.2, and the marginal probability of a nest having two or more eggs is 0.8. We can thus write

$$\Pr(N = 1) = 0.2$$
$$\Pr(N \geq 2) = 0.8$$

These are the same two probabilities that we encountered in the last chapter. The probabilities associated with the two possible values of N sum to 1, as required by the basic rule of probability. We can similarly write the marginal probabilities of a randomly chosen tree being coniferous or broad-leaved as

$$\Pr(T = \text{Conif}) = 0.75$$
$$\Pr(T = \text{Broad}) = 0.25$$

Once again, the probabilities associated with the two possible values of T sum to 1.

The second type of probability that we can find in our contingency table is **joint probability**. Joint probabilities give the probability that all of the random variables in a problem take on particular values simultaneously, or jointly. These probabilities are listed in the interior of the contingency table. For example, the probability that a nest has one egg *and* is found in a broad-leaved tree is

$$\Pr(N = 1, T = \text{Broad}) = 0.1$$

Because both random variables are taking particular values simultaneously, the order of the variables in a joint probability doesn't matter. In other words, $\Pr(N = 1, T = \text{Broad}) = \Pr(T = \text{Broad}, N = 1)$.

There are four joint probabilities in this chapter's problem, corresponding to the four cells found in the interior of the contingency table. Every nest that we observe must be associated with one combination of values of N and T. As such, these four joint probabilities also sum to 1.

The third and final type of probability requires some extra calculation, but is the key to answering the question posed in this chapter. **Conditional probability** is the probability that one random variable takes a particular value given that another random variable is already known to take a particular value. The question posed in this problem requires us to calculate the probability that a nest will have two or more eggs conditional on the tree it is in being coniferous or being broad-leaved. These conditional probabilities will tell us the type of tree that the camera crew should climb to have the greatest chance of finding a nest with two or more eggs.

Let's calculate the conditional probability first for broad-leaved trees. Returning for a moment to the counts of nests, we know that the study found 5 nests in broad-leaved trees. Of these 5 trees, 3 of them had nests containing two or more eggs. If we were to climb a broad-leaved tree, we would thus expect there to be a probability of 3/5 = 0.6 of finding a nest with two eggs in it. We can write this conditional probability as

$$\Pr(N \geq 2 \mid T = \text{Broad}) = 3/5 = 0.6$$

where the vertical bar in the probability is read as "conditional on" or "given that." We thus read the statement above as "the probability of finding two or more eggs in a nest, given that the nest is in a broad-leaved tree, is 0.6." We can calculate the equivalent conditional probability for coniferous trees as

$$\Pr(N \geq 2 \mid T = \text{Conif}) = 13/15 = 0.87$$

It appears that if the camera crew wants the highest probability of finding a nest with two or more eggs, they should climb a coniferous tree, where they have an 87% chance of finding a nest with two or more eggs. The probability of finding a nest with two or more eggs in a broad-leaved tree, by comparison, is only 60%.

The key to successfully solving this problem, and other ones like it, is recognizing precisely which kind of probability provides the answer that we are looking for. For example, we are not interested in comparing the joint probabilities of finding a nest with two or more eggs and finding a coniferous or a broad-leaved tree. This is because the problem states that we are going to choose one type of tree or the other, and we want to compare the probabilities of finding nests with two or more eggs *given* that we have chosen one particular type of tree.

Similarly, we are not interested in the conditional probability that we are in a coniferous or broad-leaved tree given that we have found a nest with two or more eggs. These probabilities are

$$\Pr(T = \text{Broad} \mid N \geq 2) = 3/16 = 0.19$$
$$\Pr(T = \text{Conif} \mid N \geq 2) = 13/16 = 0.81$$

These two conditional probabilities describe a situation in which we have already climbed a tree and found two or more eggs in the nest, and we then ask for the probability that the tree we have already climbed is coniferous or broad-leaved. The order of the two variables in conditional probabilities are thus not interchangeable. We'll see a very important and common example of mixing up this order in the next chapter.

Before moving on, there is one additional relationship that we can see in the contingency table that will be useful later on: a conditional probability can be calculated from the ratio of a joint and a marginal probability. All three types of probabilities described in this chapter are thus interrelated. For example, the conditional probability of finding a nest with two or more eggs, given that the tree is broad-leaved, can be written as

$$\Pr(N \geq 2 \mid T = \text{Broad}) = \frac{\Pr(N \geq 2, T = \text{Broad})}{\Pr(T = \text{Broad})} = \frac{3/20}{5/20} = \frac{0.15}{0.25} = \frac{3}{5} = 0.6$$

We'll make use of this relationship between conditional, joint, and marginal probabilities in the next chapter.

NEXT STEPS

Following up on the discussion of independence in the last two chapters, this chapter presents a clear example of two observations that are *not* independent, which are the type of tree a nest is found in and the number of eggs in that nest. You may recall from the last chapter that if two variables are independent, we can multiply the values of their probabilities together to get the joint probability of the two values of the two variables occurring together. In this chapter's problem, this would mean that if the variables N and T were independent, we could calculate the joint probabilities inside the contingency table by multiplying the marginal probabilities outside of the table. For example, if N and T were independent, then the joint probability $\Pr(N = 1, T = \text{Broad})$ would be equal to $\Pr(N=1) \times \Pr(T=\text{Broad}) = 0.2 \times 0.25 = 0.05$, which is not the same as the actual value of 0.1 given by the problem. The widely used χ^2 test for the independence of two variables is based on comparing a contingency table made with the assumption of independence with an actual observed contingency table.

In this particular problem, it's interesting to note the importance of simplifying the possible values of the variable N to only two: one egg and two or more eggs. If we had constructed instead a two-row by three-column contingency table, with all possible values of N, we could not have solved for the values of all six cells in the interior of the table. Had the problem instead asked which tree should be climbed to have the highest probability of finding exactly three eggs, we would not be able to provide an answer without additional information.

9 Picking Ticks Bayes's Rule

> *Over the course of several months assisting with field research, you've found and removed several ticks from your body. There is a tick-borne disease that infects about 1% of people living in this area, and you decide to take a blood test to see if you have been infected. The test is not perfect, but it is fairly accurate. The test correctly identifies the disease in every person who has it, and it correctly says that a person without the disease does not have the disease in 99% of tests. Unfortunately, your test comes back positive. What is the probability that you actually have the disease?*

This problem presents a classic case in which a careful application of the rules of probability can prevent us from making a significant error in judgment. Given the very high accuracy of the test, many people would assume that the probability that you have the tick-borne disease is something like 99%. In fact, a large proportion of medical doctors consistently give answers like this when presented with similar problems (for example, see Kremer 2014). As we will see below, this is a mistake that arises from a misunderstanding of conditional probability.

The first step in solving this problem is to recognize that the problem includes two random variables. The first random variable, which we'll call D, is whether you actually have the tick-borne disease (D = Yes) or not (D = No). The second random variable is whether your test came back positive (T = Pos) or negative (T = Neg). Just as in chapter 8, we can arrange these two random variables in a contingency table:

	D = Yes	D = No
T = Pos		
T = Neg		

Although all of the information provided in the problem is in the form of probabilities, it is easier to think through the problem if we use numbers of people instead. As we'll see later on, we will get the same answer regardless of whether we work directly with probabilities or with counts. Let's assume that we have a hypothetical population of 100,000 people living in this area, all of whom were tested for the disease. Each person may have the disease or not, and may have had a positive or negative test result. To solve the problem, we want to consider the number of people who have had a positive test result, like you, and calculate the proportion of those people who also have the disease. This is equivalent to the conditional probability $\Pr(D = \text{Yes} \mid T = \text{Pos})$.

Let's start filling in the contingency table with the numbers of people in different categories. The first piece of information that we are given is that 1% of people in this area have the disease. This means that out of 100,000 people, 1,000 have the disease and 99,000 do not. We add these numbers to the margin of the contingency table:

	D = Yes	D = No
T = Pos		
T = Neg		
	1,000	99,000

Next, we know that the test correctly identifies every person with the disease as having it. In the language of conditional probability, we can write $\Pr(T = \text{Pos} \mid D = \text{Yes}) = 1$. This means that all 1,000 people with the disease test positive, and none test negative.

	D = Yes	D = No
T = Pos	1,000	
T = Neg	0	
	1,000	99,000

Finally, we know that the test incorrectly comes back positive for 1% of people without the disease. In other words, $\Pr(T = \text{Pos} \mid D = \text{No}) = 0.01$. This means that out of the 99,000 people without the disease, 990 will test positive and 98,010 will test negative.

	D = Yes	D = No
T = Pos	1,000	990
T = Neg	0	98,010
	1,000	99,000

We now have all four interior cells filled in, which allows us to fill in the margin on the right of the table, which reports the total number of positive and negative tests in this group of 100,000 people:

	D = Yes	D = No	
T = Pos	1,000	990	1,990
T = Neg	0	98,010	98,010
	1,000	99,000	

We're now ready to determine the probability that you have the tick-borne disease, given that you tested positive, $\Pr(D = \text{Yes} \mid T = \text{Pos})$. From the table, we can see that there were 1,990 positive tests, out of which 1,000 people actually had the disease. Thus we can calculate

$$\Pr(D = \text{Yes} \mid T = \text{Pos}) = \frac{1{,}000}{1{,}990} = 0.502$$

Despite the apparent accuracy of the test, you thus have only a 50% chance of having the disease, given that you tested positive!

This number may be quite different from your intuition that this probability should have been much higher. This mistake stems from confusion about the conditional probabilities that we are trying to calculate. Specifically, the problem tells us that $\Pr(T = \text{Pos} \mid D = \text{Yes}) = 1$ and $\Pr(T = \text{Neg} \mid D = \text{No}) = 0.99$. These numbers seem to suggest that we should be very confident in the conclusions of the test. So why, then, is our probability of interest, $\Pr(D = \text{Yes} \mid T = \text{Pos})$, so unintuitively low?

The basic explanation, in words, goes like this. In reality, only 1% of the population has the disease and will test positive for it. However, 1% of the remaining 99% of the population *without* the disease will also test positive for it. There is thus an almost identical number of people in the "tested positive" group who tested positive because they had the disease and who tested positive incorrectly even though they don't have the disease. As a result, only about half of the people who test positive actually have the disease.

We can gain even better intuition by playing around with different values for the test accuracy and the background rate of the disease. Rather than rewriting our contingency table over and over again, let's instead take a moment to consider how we could calculate our probability of interest, $\Pr(D = \text{Yes} \mid T = \text{Pos})$, directly from the probabilities given in the problem.

As discussed in the last chapter, conditional probability can be written as the ratio of a joint probability and a marginal probability. In general, if A and B stand for two random variables and X and Y stand for some values that these variables can take, we can write

$$\Pr(A = X \mid B = Y) = \frac{\Pr(A = X, B = Y)}{\Pr(B = Y)}$$

We actually used this formula implicitly in determining the answer to this problem:

$$\Pr(D = \text{Yes} \mid T = \text{Pos}) = \frac{\Pr(D = \text{Yes}, T = \text{Pos})}{\Pr(T = \text{Pos})} = \frac{1{,}000 / 100{,}000}{1{,}990 / 100{,}000}$$

$$= \frac{1{,}000}{1{,}990} = 0.502$$

Let's now consider how we would calculate the "reverse" of this conditional probability, which would instead reflect the probability of testing positive given that you have the disease. We can write this probability as

$$\Pr(T = \text{Pos} \mid D = \text{Yes}) = \frac{\Pr(T = \text{Pos}, D = \text{Yes})}{\Pr(D = \text{Yes})} = \frac{1{,}000 / 100{,}000}{1{,}000 / 100{,}000} = 1$$

This result confirms what we already knew, which is that the test always comes back positive if a person has the disease.

You might notice something interesting about the two equations above: both equations contain the joint probability of testing positive and having the disease. As described in the last chapter, the order of the variables in a joint probability does not matter. We can thus rearrange these two equations and set the two joint probabilities equal to each other, which gives us

$$\Pr(D = \text{Yes} \mid T = \text{Pos}) = \frac{\Pr(T = \text{Pos} \mid D = \text{Yes})\Pr(D = \text{Yes})}{\Pr(T = \text{Pos})}$$

This equation effectively provides the formula for exchanging the order of variables in a conditional probability statement, from $\Pr(T = \text{Pos} \mid D = \text{Yes})$ to $\Pr(D = \text{Yes} \mid T = \text{Pos})$ in this case. The equation above comes up frequently throughout probability and statistics and is known as **Bayes's rule**.

Let's use Bayes's rule to help us get some additional intuition as to why our initial guess as to the probability that you had the disease given that you tested positive may have been so far off. Notice that this probability is exactly what is written on the left-hand side of this equation, so the right-hand side of the equation provides a means to calculate it. We know that $\Pr(T = \text{Pos} \mid D = \text{Yes}) = 1$, as described above, and we know that $\Pr(D = \text{Yes}) = 0.01$, since 1% of the people in this area have the disease.

Calculating the denominator, $\Pr(T = \text{Pos})$, will require a bit more thought. A person can test positive in two ways. First, with some probability, they can have the disease and test positive. Second, with some probability, they can *not* have the disease but test positive anyway. The overall probability of testing positive is the combined probability of these two paths to receiving a positive test. We can express this idea, using our probability formulas, as

$$\Pr(T = \text{Pos}) = \Pr(D = \text{Yes})\Pr(T = \text{Pos} \mid D = \text{Yes}) + \Pr(D = \text{No})\Pr(T = \text{Pos} \mid D = \text{No})$$

This method of calculating a marginal probability from conditional probabilities is known as the **law of total probability**.

Substituting this long equation into the denominator of Bayes's rule gives the expanded equation

$$\Pr(D = \text{Yes} \mid T = \text{Pos}) = \frac{\Pr(T = \text{Pos} \mid D = \text{Yes})\Pr(D = \text{Yes})}{\Pr(D = \text{Yes})\Pr(T = \text{Pos} \mid D = \text{Yes}) + \Pr(D = \text{No})\Pr(T = \text{Pos} \mid D = \text{No})}$$

Although it's a bit unwieldy, this equation is also very useful. We can now substitute the probabilities that were given in the problem into this equation, which gives us

$$\Pr(D = \text{Yes} \mid T = \text{Pos}) = \frac{1 \times 0.01}{0.01 \times 1 + 0.99 \times 0.01} = 0.502$$

By playing around with this faster method of calculation, we can see that the background rate of the disease in the population plays a dramatic role in our answer to the problem. For example, if 10% of the population had the disease, rather than 1%, we would find instead that

$$\Pr(D = \text{Yes} \mid T = \text{Pos}) = \frac{1 \times 0.1}{0.1 \times 1 + 0.9 \times 0.01} = 0.92$$

which might more closely match our intuition. In short, when a disease is rare, very few people actually have it and a very large number of people do not. Even if a fairly small percentage of people without the disease mistakenly test positive, that will still lead to a large fraction of the positive tests coming from these false positives, rather than from people who actually have the disease. It turns out that this exact situation is quite common in medical diagnostics.

We can also explore a variety of other possibilities using the equation above. Imagine, for example, that the test correctly said that 95% of people without the disease did not have the disease, which still seems like a fairly accurate test. In this case, the probability of having the disease given a positive test would be only 17%!

NEXT STEPS

Misunderstandings about conditional probabilities are particularly important to keep in mind whenever a person is diagnosed with a rare disease using an imperfect test. However, this problem is not restricted to medicine. In other fields, it is known as the "base rate fallacy" or the "prosecutor's fallacy." Consider, for example, a criminal case in which a defendant was picked randomly from a list of 100 potential suspects and found to have a fingerprint that matches the one left at a crime scene. The fingerprint-matching algorithm has only a 1% chance of mistakenly returning a match if the defendant did not, in fact, leave the fingerprint. In other words, Pr(Fingerprint = Match | Defendant = Innocent) = 0.01, which might sound like a very strong basis for a guilty verdict. However, what the jury actually needs to determine is Pr(Defendant = Innocent | Fingerprint = Match). These two conditional probabilities are not the same, but are instead related by Bayes's rule. Using the equation for Bayes's rule above, and assuming that a guilty person always leaves a matching fingerprint and that the guilty person is in the group of 100 suspects, you should be able to find that Pr(Defendant = Innocent | Fingerprint = Match) also, in fact, is just about 0.5. Intuitively, there's roughly an equal chance that the matching fingerprint was correctly matched from the one guilty suspect and incorrectly matched from one of the 99 innocent suspects.

The numbers in this problem are based on an example of testing for Lyme disease given by Kling et al. (2015). To simplify the problem slightly, we've assumed here that the test is perfectly accurate in detecting the disease when it is present.

As a final note, the use of Bayes's rule is part of both frequentist and Bayesian probability and statistics. Bayesian statistics does get its name, however, from its very deep reliance on the use of Bayes's rule.

10 **Rabbit Rates** Probability Distributions

A population of rabbits lives in the forest surrounding the eagle nest that was eventually located by the camera crew. Each day, the rabbit living closest to the nest has a 5% chance of being caught by the eagles. What is the probability that this rabbit will be eaten on days 0, 1, 10, or 100? What is the probability that it will have been eaten on or before day 100?

The first part of this problem asks for the probability that this unfortunate rabbit is eaten on three particular days. At first, the answer might seem simple. If the rabbit has a 5% chance of being eaten each day, then we might initially think that the probability that the rabbit ends up being eaten on any day t would be 5%. Reflecting on this for a moment, however, we can see that this seemingly simple answer can't be right. If we were to add up a 5% probability of being eaten on each day up until day 100, we would conclude that the rabbit had a 500% chance of being eaten before day 100. The mistake in our initial suggestion comes from a failure to consider that a rabbit must still be *alive* at the start of some day t in order for it to be eaten on that particular day. In other words, it can't be eaten on day t if it has already been eaten on some day before t.

Let's put this idea into an equation. We'll define the parameter $p = 0.05$ as the probability that the rabbit is eaten each day given that it is alive at the start of that day. We'll start with the rabbit alive and begin projecting the probability that the rabbit is eaten on day t starting with day 0, when $t = 0$. This is a bit different from our models in part 1, where we started with our initial population size at $t = 0$ and began our projections with time step $t = 1$. Starting with $t = 0$ instead, however, will eventually lead us to a well-known equation that we'll use again in chapter 18.

We'll thus begin with the probability that the rabbit is eaten on day 0, at $t = 0$. Given that the rabbit starts day 0 alive, this probability is logically equal to $p = 0.05$. If we use the random variable E to represent the day on which the rabbit is eaten, we can write

$$\Pr(E = 0) = p = 0.05$$

for the probability that the rabbit is eaten on day 0.

Next, we can consider the probability that the rabbit is eaten the next day, at $t = 1$. We've already decided above that this probability can't once again be $p = 0.05$. Instead, for the rabbit to be eaten on day 1, it first has to *not* be eaten on day 0 and then be eaten on day 1. Since a rabbit can only be eaten or not eaten on a day, the probability of not being eaten on a day must be $1 - p = 0.95$. The joint probability that the rabbit is not eaten on day 0 and is eaten on day 1 is thus

$$\Pr(E = 1) = (1 - p)p = 0.0475$$

where $(1 - p)$ represents "not being eaten" on day 0 and p represents "being eaten" on day 1. Notice that in multiplying these two probabilities together, we are assuming a type of independence between the days, in that the probability of being eaten or not on day 1 is not affected by the fact that the rabbit was not eaten on day 0.

If we continue this same logic for additional days, we find that

$$\Pr(E = 2) = (1 - p)(1 - p)p$$
$$\Pr(E = 3) = (1 - p)(1 - p)(1 - p)p$$
$$\Pr(E = 4) = (1 - p)(1 - p)(1 - p)(1 - p)p$$

and so on.

We can see that a general pattern is emerging in this sequence of equations. It appears that we can calculate the probability of being eaten on any day t by multiplying the probability of not being eaten a total of t times up until that day by the probability of finally being eaten on that day. This means that we can calculate the probability of being eaten on any day t as the probability $(1 - p)$ multiplied together t times, multiplied by one final p. This leads us to infer the general equation for the probability of being eaten on day t as

$$\Pr(E = t) = (1 - p)^t p$$

This equation defines a probability distribution, specifically known as the **geometric distribution**, for the random variable E. As described in chapter 5, a probability distribution is simply a list of probabilities associated with all of the possible values of a random variable, where all of the probabilities add up to 1. In this problem, the rabbit can, in principle, be eaten on any day from day 0 up to day infinity. It turns out that the equation above is cleverly designed such that the sum of the probabilities of being eaten on each day from 0 to infinity will always add up to 1, so long as the value of p is between 0 and 1. Figure 10.1 shows a plot of this geometric distribution for each day up to $t = 100$.

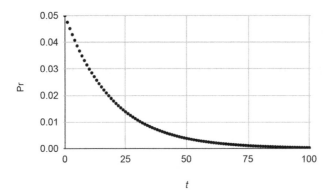

Figure 10.1. Probability of the rabbit being eaten on each day from 0 through 100. These probabilities are described by a geometric distribution.

In a new Google Sheet, type the variable names `t` and `Pr` in Cells A1 and B1. In Cell D1, enter the variable name `p`, and in Cell D2, enter the number `0.05`. Starting in Cell A2, enter the numbers from 0 to 100 in successive rows. In Cell B2, enter the formula `=(1-$D$2)^(A2)*$D$2`. Highlight this cell and drag it down to Row 102.

Looking at the equation for the geometric distribution, you might notice that we've seen something similar before. In fact, we met this same equation back in chapter 2, where we derived the equation for exponential, or geometric, growth of a population. In both cases, we calculate the value of a variable by multiplying the last value by a constant parameter. In chapter 2, this parameter was R, the population growth rate, while here it is the probability $(1 - p)$. That explains why this probability distribution is known as the geometric distribution.

With the equation for the geometric distribution, we can easily calculate the probability that our rabbit is eaten on any particular day. To answer the original question, $\Pr(E = 0) = 0.05$, $\Pr(E = 1) = 0.0475$, $\Pr(E = 10) = 0.03$, and $\Pr(E = 100) = 0.0003$. In general, the probability of the rabbit being eaten on a particular day decreases as time goes on. This is because as t increases, there is a greater and greater chance that the rabbit has already been eaten, thus making it unavailable to be eaten on day t.

Instead of asking about the probability that the rabbit is eaten on a particular day, the second part of the problem asks for the probability that the rabbit has been eaten *by* a particular day—for example, on day 100 or before. Logically, the probability that the rabbit is eaten on day 100 or before should be the probability that the rabbit is eaten on day 0, plus the probability that it's eaten on day 1, and so on until day 100. Since a rabbit can be eaten on only one day, we can add these probabilities together to determine the probability that it is eaten on any one of these days. We can write this probability as

$$\Pr(E \leq 100) = \Pr(E = 0) + \Pr(E = 1) + \Pr(E = 2) + \ldots + \Pr(E = 100)$$

This turns into the equation

$$\Pr(E \leq 100) = p + (1-p)^1 p + (1-p)^2 p + \ldots + (1-p)^{100} p$$

Although we could calculate and add up all of these numbers, there's actually an easier way to calculate this probability. In addition to being the sum of the terms above, the probability that the rabbit is eaten on or before day 100 is also equal to one minus the probability that the rabbit is still alive on day 101. The probability that the rabbit survives any one day is still $(1-p)$, and thus the probability that it survives 101 days in a row, including day 0, to reach $t = 101$ alive is $(1-p)^{101}$. The probability that the rabbit is eaten at some point on or before day 100 is thus $1-(1-p)^{101}$. More generally, for any day t, the probability that the rabbit is eaten on or before day t is

$$\Pr(E \leq t) = 1 - (1-p)^{t+1}$$

This equation provides us with the cumulative probability that the rabbit is eaten on one of the days up to and including day t. As such, it is known as the **cumulative distribution function**, or CDF, for this geometric distribution. We can use this equation to calculate, for example, the probability that the rabbit is eaten on or before day 100, as $1 - 0.95^{101} = 0.9944$. After 100 days, there is a 99.4% chance the rabbit has already been eaten.

NEXT STEPS

This chapter introduced our first probability distribution, the geometric distribution, that is defined by an equation rather than given simply as a list of numbers. As described above, we derived a form of the geometric distribution with support from 0 to infinity. It's worth knowing that this is one of two common forms of the geometric distribution in widespread use. The other form is defined by the equation $\Pr(E = t) = (1-p)^{t-1} p$, where t now runs from 1 to infinity instead of 0 to infinity. We focused on the version with support from 0 to infinity because we will use this same distribution in part 4.

In part 1, we noted that geometric population growth, which proceeds in discrete time steps such as days, is a discrete-time analogue to the continuous-time exponential growth equation. Similarly, the geometric distribution is the discrete-support analogue to the continuous-support exponential distribution. We won't discuss any continuous probability distributions in this book, although you can find a discussion of these in any book on probability and statistics.

Part III **Modeling Multiple States**

11 Introducing Matrix Models

In parts 1 and 2 of this book, we solved problems that required us to estimate and track the value of one or more variables over time. The problems in part 3 of this book will continue exploring these ideas by presenting the framework of **matrix models**, which provide an efficient way of solving problems involving multiple variables.

The most important thing to recognize about matrix models is that they can be understood as a multiple-variable extension of the simple exponential growth model that we met in chapter 2. Returning to the problem from that chapter, we decided that we could model the growth of a patch of duckweed in a pond over time using the difference equation

$$N_{t+1} = RN_t$$

As described in that chapter, N_t is the area of the duckweed patch at time t, and R is the multiple by which the population increases in a single time step. In that problem, we had $R = 1.4$, so that each day the area of the patch grew by 40%.

Let's imagine now, however, that we want to examine two adjacent plots in the pond, a northern plot and a southern plot. While the area of duckweed in each plot grows by the same 40% in each time step, imagine that some amount of duckweed floats from the northern plot to the southern plot each day. If after the duckweed area in each plot grows each day, 1/7 of the duckweed in the northern plot floats to the southern plot that same day, how will the area covered by duckweed change in the two plots over time?

In this modified problem, we are interested in tracking the area of duckweed in two plots in the pond, which means we'll need to track change in two different variables. We'll call these variables N_t for the area of the northern duckweed plot and S_t for the area of the southern duckweed plot. If the two plots were not connected, we could use the equations

$$N_{t+1} = 1.4N_t$$

and

$$S_{t+1} = 1.4S_t$$

to describe the change in duckweed area in each plot. However, in this new scenario, 1/7 of the duckweed area in the northern plot floats to the southern plot in each time step. This means that the duckweed area in the northern plot increases to only $1.4N_t \times 6/7 = 1.2N_t$ in each time step, while $1.4N_t \times 1/7 = 0.2N_t$ is added to the southern duckweed plot. This gives us the linked equations

$$N_{t+1} = 1.2N_t$$
$$S_{t+1} = 1.4S_t + 0.2N_t$$

We could proceed with projecting these two equations over time using the same approaches that we learned in part 1.

Notice, however, that the two equations above have a special form. Both equations closely resemble the original exponential growth equation for the single pond, in that the area covered by duckweed in the next time step is calculated by taking the area in each plot in the previous time step and multiplying it by a constant. In the world of ecological modeling, equations of this type are known as **linear equations**. This is not because the population size increases over time following a straight line, which it definitely does not. Rather, it's because the population size in the *next time step*, or equivalently, the change in population between time steps, can be calculated by multiplying the population in the *current time step* by a constant. The relationship between the variables N_{t+1} and N_t is thus described by a line.

When we have a set of equations that takes this special form, it turns out that we can express and solve these equations using the framework of matrix modeling. Using matrix models requires understanding just a little bit of **linear algebra**. Fortunately, we'll need to know only three things about linear algebra to understand all of the problems in part 3 of this book, as well as many of the matrix models that you'll encounter beyond this book.

First, we'll need to understand the concept of a **vector**. In the simplest possible terms, we can think of a vector as simply two or more numbers joined together as one variable. In this book, we'll be working with column vectors, which are made up of two or more numbers stacked on top of one another. For example, we can define a vector named $\mathbf{n}_t$ as

$$\mathbf{n}_t = \begin{pmatrix} a \\ b \end{pmatrix}$$

where a and b can be any numbers that we'd like. Anticipating how we will use this vector later, we can use the values of the variables N_t and S_t as the numbers a and b. This gives us

$$\mathbf{n}_t = \begin{pmatrix} N_t \\ S_t \end{pmatrix}$$

Notice that we write the name of the vector $\mathbf{n}_t$ using boldface text to make it clear that this variable is a vector and not just a single number.

Second, we'll need to understand the concept of a **matrix**. Like a vector, a matrix contains multiple numbers, but this time in two dimensions. A two-by-two matrix $\mathbf{A}$, for example, would be written as

$$\mathbf{A} = \begin{pmatrix} c & d \\ e & f \end{pmatrix}$$

where the lowercase letters can be any numbers that we'd like. Once again anticipating how we'll use this matrix later, we'll fill it in with the numbers from our expanded duckweed problem, so that we have

$$\mathbf{A} = \begin{pmatrix} 1.2 & 0 \\ 0.2 & 1.4 \end{pmatrix}$$

We'll see in just a moment why these particular numbers were placed in these particular locations. We write the name of a matrix, like the name of a vector, in boldface text to indicate that it contains more than one number.

Third, and finally, we'll need to know the rule for **matrix multiplication**, which in our case will be the method for multiplying our matrix ($\mathbf{A}$) by our column vector ($\mathbf{n}_t$). In contrast to "regular" multiplication, the order of the matrix and the vector does matter in matrix multiplication. In the problems in this chapter, we'll be concerned with calculating the value of the product $\mathbf{A}\mathbf{n}_t$, which is equal to

$$\mathbf{A}\mathbf{n}_t = \begin{pmatrix} 1.2 & 0 \\ 0.2 & 1.4 \end{pmatrix} \begin{pmatrix} N_t \\ S_t \end{pmatrix}$$

The rule for multiplying a matrix by a column vector is actually fairly straightforward. Imagine taking the vector on the right and "tipping it over" onto its side, then "laying it on top of" each of the rows in the matrix. We then multiply the numbers that are "on top of" each other in each row and add the products together. When we do this for each of the rows of the matrix, we create a new vector with the same number of rows as matrix A.

When we perform this operation for A and n_t, we create a new vector n_{t+1} that has the values

$$n_{t+1} = An_t = \begin{pmatrix} 1.2 & 0 \\ 0.2 & 1.4 \end{pmatrix} \begin{pmatrix} N_t \\ S_t \end{pmatrix} = \begin{pmatrix} 1.2N_t + 0S_t \\ 0.2N_t + 1.4S_t \end{pmatrix}$$

Notice that the equations in the top and bottom rows of our new vector n_{t+1} are exactly equal to the equations used to calculate N_{t+1} and S_{t+1}! We thus can see that

$$n_{t+1} = \begin{pmatrix} N_{t+1} \\ S_{t+1} \end{pmatrix}$$

and that the equation $n_{t+1} = An_t$ gives us with a compact way of projecting the area of the duckweed patches in *both* the northern and southern plots using a single equation.

Although it is arguable whether using a matrix model is helpful for a simple example like this one, the next four chapters will present problems in which it is not only efficient to use a matrix model, but in which the use of the matrix model also aids our understanding. Chapters 12 and 13 present problems similar to those in part 1, where we wish to track the sizes of multiple interacting populations over time. Chapters 14 and 15 will expand on the problems in part 2 by introducing the idea of tracking changes in probabilities over time.

12 **Imagine All the Beetles** Age-Structured Models

> *A ranger at a nearby state park has recently trapped several beetles of a non-native species that had not previously been seen at the park. A study in the beetle's native range, where the insect is a crop pest, found that the beetles appear to live up to three years in the wild. First-year beetles have a 30% chance of surviving into their second year, and second-year beetles have a 50% chance of surviving into their third year. Second-year females produce 2 surviving offspring on average, and third-year females produce 10 surviving offspring on average. The ranger fears that there could be as many as a thousand first- and second-year individuals currently in the park. How many beetles could be present in the park in 20 years if no control actions are taken? How much would it help to set traps that catch 5% of the second- and third-year adults that would have otherwise survived each year?*

At first glance, this problem appears very similar to the population growth problems from part 1, in which we wish to project the population of a single species into the future. Although there is only one beetle species in this problem, there appear to be important differences between beetles in their first year, second year, and third year of life. Because of these differences, we'll want to use three separate variables in order to track the population of beetles in each of these three age classes separately. This type of model is known as an **age-structured model**.

Before we can get started, however, we need to consider a complication that we have not encountered in any of our previous population models. It's not stated in the problem, but it's probably safe to assume that this beetle species reproduces sexually, which means that there are male and female beetles in the population. Male beetles, however, cannot give birth to offspring. To address this complication, a simple approach is to assume that half of the population consists of males and half consists of females at all time periods. With this assumption, we can see that while the average number of offspring produced by a second-year female is 2, the average number produced by a second-year *individual* is 1, since half of the second-year individuals do not produce offspring. Similarly, the average number of offspring per third-year individual is 5, instead of 10. A more exact approach would be to build separate and linked projections for the male and female populations, which is something that you may be interested in trying after working through the equations below.

With that assumption in hand, we can get started by setting up a system of three difference equations, one for each of the three possible age classes. If we use the variables F, S, and T to represent the numbers of beetles in their first, second, and third years, respectively, we can write down the following equations to describe how the size of each age class changes over time:

$$F_{t+1} = 1S_t + 5T_t$$
$$S_{t+1} = 0.3F_t$$
$$T_{t+1} = 0.5S_t$$

These equations express all of the information given in the problem. To start, the number of first-year individuals at time $t + 1$ is determined by the number of individuals that are born from the existing second-year and third-year individuals at time t. As described above, these numbers are one-half of the numbers given in the problem, which are for female beetles only. The number of individuals in the second-year age class at $t + 1$ is then equal to 30% of those that were in the first-year age class at time t, since surviving the first year is the only way that a second-year individual can be "born." Similarly, the number of individuals in the third-year age class at time $t + 1$ is 50% of those that were in the second-year age class at time t.

Next, we'll use these equations to set up a matrix model of the type described in the last chapter. Let's start by expanding our system of three equations a bit so that the right-hand side of each equation includes all three of our variables in the same order. The set of equations now looks like this:

$$F_{t+1} = 0F_t + 1S_t + 5T_t$$
$$S_{t+1} = 0.3F_t + 0S_t + 0T_t$$
$$T_{t+1} = 0F_t + 0.5S_t + 0T_t$$

This expansion shows us how to fill in the values of our **projection matrix**, which describes, in matrix form, how a certain number of individuals in each class at time t produces a certain number of individuals in each class at time $t + 1$. Such a projection matrix is commonly given the name A. For our problem, the matrix looks like this:

$$A = \begin{pmatrix} 0 & 1 & 5 \\ 0.3 & 0 & 0 \\ 0 & 0.5 & 0 \end{pmatrix}$$

Now that we have our projection matrix, we'll need a vector of the starting population of beetles at $t = 0$. The problem states that the ranger fears that there could be up to a thousand first- and second-year individuals currently in the park, and we'll use that number as a type of worst-case scenario. Since we don't know specifically how these individuals might be distributed across the first- and second-year age classes, we'll assume that half are in their first year and half in their second year. Our vector of population sizes for each age class at $t = 0$, which we'll name n_0, is thus

$$n_0 = \begin{pmatrix} 500 \\ 500 \\ 0 \end{pmatrix}$$

In this vector, the top row contains the value of F_0, the middle row S_0, and the bottom row T_0.

As described in the last chapter, we can then project the population of each age class for $t = 1$, the next time step, using the equation

$$n_1 = An_0$$

Let's perform this multiplication by hand using the rules described in the last chapter. The first row in the matrix A contains the numbers 0, 1, and 5. The vector n_0 contains the numbers 500, 500, and 0 arranged vertically in a column. If we take the column, tip it over, lay it on top of the first row of A, and multiply the numbers that end up on top of each other, we get 0×500, 1×500, and 5×0. Adding these up gives us $0 + 500 + 0 = 500$. If we repeat the same procedure for the second and third rows, we get the results 150 and 250. Placing those three results in a new vector n_1, we find that

$$n_1 = \begin{pmatrix} 500 \\ 150 \\ 250 \end{pmatrix}$$

Notice that the three values in the vector n_1 are exactly the values of F_1, S_1, and T_1 that we could have calculated using the system of three separate

Imagine All the Beetles: Age-Structured Models

equations at the start of this chapter. In fact, the rules of matrix multiplication are specifically designed to give the exact same result you would get by projecting that system of linear equations one time step into the future. This matrix multiplication procedure can be repeated over and over again, like our methods from part 1, to project the population of all three age classes into the future, with

$$n_1 = An_0$$
$$n_2 = An_1$$
$$n_3 = An_2$$

and so on for as long as we'd like.

Rather than calculating each vector n_t by hand, we can use matrix multiplication functions to help us easily project the population of each age class into the future. This method allows us to calculate the total population size after 20 years as 1,661 beetles. This is an increase of only 661 beetles over the number currently thought to be present, which seems like a fairly small increase for a 20-year period. In this final population, 1,143 individuals are first-years, 355 individuals are second-years, and 163 individuals are third-years.

> In a new Google Sheet, enter the nine values in the matrix *A* in the first three rows and first three columns of the sheet (i.e., Cells A1 to C3). Then, in Cells A5 through A8, enter the variable names t, Ft, St, and Tt. In Cell B5, enter 0, and increase this number to 20 in successive cells to the right in this same row. In Cells B6 through B8, enter the values 500, 500, and 0, corresponding to the initial population sizes in this problem. To project the population using matrix multiplication, enter the formula =MMULT($A$1:$C$3,B6:B8) in Cell C6, then drag the corner of this cell down to Cell C8 (if the numbers in Cells C7 and C8 do not appear automatically). You should see the numbers 500, 150, and 250 in Cells C6 to C8, which are the values of the vector n_1. Highlight Cells C6 to C8 and drag these to the right to Column V to project the population to year 20. Adding up the populations of all three age classes in any year gives the total population size in that year.

The relatively slow growth of this beetle population would undoubtedly be something of a relief to the park manager. If the figures from the published study do indeed apply to the beetles in the park, and our initial assumption about the number of beetles in each age class is correct, it appears that there will not be a major outbreak of these beetles any time soon. Still, the manager might wonder about the effect of setting out traps, which could catch 5% of the beetles in the second- and third-year age classes that would otherwise have survived. This doesn't seem like a large percentage, but we can investigate the effect of the traps by modifying matrix A to

$$A = \begin{pmatrix} 0 & 1 & 5 \\ 0.25 & 0 & 0 \\ 0 & 0.45 & 0 \end{pmatrix}$$

which reflects a decrease in the percentage of first-year beetles that survive to their second year and the percentage of second-year beetles that survive to their third year. Performing the same analysis using this new matrix, we find that the beetle population is now only 259 individuals after 20 years. Increasing adult mortality with the traps even by this small amount has thus caused the population to decline instead of grow.

Before we move on from this problem, there are three interesting points to notice about this model and the results that we've calculated. To help illustrate these points, figure 12.1 shows the projected beetle population under the no-trap scenario for 150 years into the future. All three age classes are shown, along with the total number of beetles across all age classes.

The first point relates to the rate of increase of the beetle population from year to year. In the first few years, we can see that the number of beetles in each age class seems to jump around quite a bit, with the population in each class sometimes increasing and sometimes decreasing. However, around year 20, this jumpiness seems to settle down, and the number of beetles in each age class increases smoothly into the future.

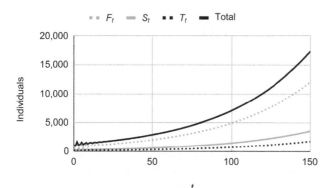

Figure 12.1. Projected populations of the first-year (F_t), second-year (S_t), and third-year (T_t) age classes for the invasive beetle under the no-trap scenario.

By about year 50, it turns out that the year-to-year change in the total population size and in all of the separate age classes can be described by a single number, λ, which in this matrix model has a value of 1.018. In other words, we can eventually project the population as a whole or the size of any age class just by calculating

$$n_{t+1} = \lambda n_t$$

where the variables n_t and n_{t+1} are single numbers, not vectors or matrices.

The second point is that we've actually already seen the equation above back in chapter 2, where we discussed the exponential growth model. In fact, matrix models lead to a kind of multi-variable exponential growth, in which all classes of the model eventually reach a point of exponential growth or decline. It should be easy to see from the equation above that if λ is greater than 1, the population will grow indefinitely, while if λ is less than 1, the population will decline to zero. Adding the traps to increase beetle mortality, in fact, reduces λ to 0.93, which will eventually eliminate the population of beetles from the park if no additional individuals are introduced.

The third point relates to the idea of the populations in each age class eventually forming the smooth curves seen after year 20 in figure 12.1. By this time, when the value of λ stabilizes, it turns out that the proportion of individuals in each age class stabilizes as well. In other words, even as the population size grows or shrinks going forward, the proportion of individuals in each of the three age classes remains constant. These proportions are known as the **stable age distribution** for the population, which is often represented by a vector named **w**. In the no-trap case, we find that after year 50 or so,

$$\mathbf{w} = \begin{pmatrix} 0.7 \\ 0.2 \\ 0.1 \end{pmatrix}$$

so that 70% of the beetles are in their first year, 20% are in their second year, and 10% are in their third year.

NEXT STEPS

The parameter λ in this chapter is closely analogous to the reproductive factor R from chapter 2. In matrix modeling, λ has a special name, which is the **dominant eigenvalue** of the matrix A. The dominant eigenvalue is the parameter that determines the rate at which the population, and each of the age classes, increases or decreases over time in the long term once the stable age distribution is reached. Similarly, the stable age distribution w also has a special name, which is the **right eigenvector** of the matrix A that is associated with the dominant eigenvalue. Very usefully, both λ and w can be calculated directly from the matrix A, without the need to project the model forward in time. See Caswell (2006) for details on this calculation and many other related details of matrix modeling. It's also worth noting that there are some projection matrices for which the population does not end up growing smoothly and exponentially with a stable age distribution. Caswell (2006) also provides an extensive discussion of how to identify these situations.

The projection matrix used in this chapter is the same as one analyzed by Caswell (2006, example 2.1). Leslie (1945) describes the use of a similar matrix model for analyzing an imaginary beetle population.

13 The Road to Succession Transition Matrices

> *A group of citizen scientists has recently surveyed a small patch of 100 trees in a local park. The survey found 1 beech, 20 maples, and 79 trees of other species in the patch. Under each adult tree, the group counted the number of saplings of each species. The proportions of sapling species found under each species of adult tree are shown in table 13.1. If we assume that each adult tree will be replaced randomly by one of the saplings under it, how will the forest composition change, if at all, in the future?*

This problem poses a question about **succession**, the process by which a community changes predictably over time following an initial disturbance. In a temperate forest like this one, the relative proportions of different tree species are often seen to change gradually over time, eventually reaching a state with a more or less stable balance of various species. To predict what this stable balance might eventually look like, we'll use a variant of the matrix models that we introduced in chapter 12. The most important difference between this problem and the one in chapter 12 is that here we will assume that the total number of adult trees remains constant at 100 individuals for all times t. Only the relative *proportions* of trees of each species change over time.

We'll begin by using the information in table 13.1 to set up a system of three difference equations that describes how the number of trees of each of the three species changes during a single cycle of adult tree replacement. Of course, in reality, not every tree in the forest dies and gets replaced by a sapling simultaneously, but it turns out that the relative proportion of individuals of each species in the eventual stable state is not affected by this assumption.

If we use the variables B, M, and O to represent beeches, maples, and other species, respectively, we can write the three difference equations describing the change in the number of trees of each species as

$$B_{t+1} = 0.8B_t + 0.25M_t + 0.10O_t$$
$$M_{t+1} = 0.1B_t + 0.15M_t + 0.25O_t$$
$$O_{t+1} = 0.1B_t + 0.6M_t + 0.65O_t$$

The first of these three equations, for example, states that the number of beech trees in the forest in the next generation is equal to 80% of the

Table 13.1 Proportions of sapling species found under each adult tree species

Sapling species	Adult tree species		
	Beech	Maple	Other
Beech	0.8	0.25	0.1
Maple	0.1	0.15	0.25
Other	0.1	0.6	0.65

beech trees in the current generation, plus 25% of the maple trees in the current generation, plus 10% of the other species of trees in the current generation. These numbers are simply the percentages of beech saplings under the adult trees of each of those three species. If each adult is replaced by a randomly chosen sapling under it, then these sapling percentages will equal the percentages of adult trees that transition from each species to beech.

We can see that these three equations could be placed in a matrix, A, very similar to the population projection matrix from the last chapter. In this case, however, A describes the transitions for a fixed number of adult trees between species groups. Since the matrix A describes transitions of a fixed number of individuals between states, it is known as a **transition matrix**. The values in this transition matrix are given by

$$A = \begin{pmatrix} 0.8 & 0.25 & 0.1 \\ 0.1 & 0.15 & 0.25 \\ 0.1 & 0.6 & 0.65 \end{pmatrix}$$

The first row of this matrix, as we previously described, gives the percentage of adult beech, adult maple, and adult trees of other species that will transition in the next generation to become beech trees. Conversely, the first column of this matrix gives the percentage of newly born beech, maple, and other trees that will emerge in the next generation from the current set of adult beech trees.

At present, the number of adult trees of the three species can be organized in the vector n_0,

$$n_0 = \begin{pmatrix} 1 \\ 20 \\ 79 \end{pmatrix}$$

which gives, from top to bottom, the number of beech, maple, and other trees currently in the forest patch.

Just as in the last chapter, we can see that by multiplying the transition matrix A by the vector n_0, we can project the number of beech, maple, and other trees in the patch at time $t = 1$, one generation later. This gives exactly the same result that we would get if we performed this calculation using the three separate difference equations that we wrote down earlier. Using matrix multiplication, we find the result

$$n_1 = \begin{pmatrix} 13.7 \\ 22.85 \\ 63.45 \end{pmatrix}$$

The three numbers in the vector n_1 sum to 100, as do the three numbers in the vector n_0. This will be true for all times t that we might examine in the future.

> The instructions from chapter 12 can be used here in a new Google Sheet to project the forest community composition into the future. Just replace the values of the matrix A and the vector n_0 with the ones in this chapter.

When we project the tree community composition into the future, we find that after about 15 generations, the community reaches a relatively steady state with about 42 beeches, 17 maples, and 41 trees of other species. These proportions are fairly different from those in the current community. This future **stationary state** of the forest community is analogous to the stable age distribution from chapter 12 and is calculated in the same manner. Given that the total size of the forest is neither increasing or decreasing over time, it should not be surprising that the value of λ for this transition matrix is equal to 1. The same is true for any transition matrix.

The fact that the number of trees neither increases or decreases over time is related to a special feature of the transition matrix A, which is that the values in each column of the matrix sum to 1. Multiplying any transition matrix with this property by a column vector will give a new column vector, and the sum of all the numbers within each column vector will be the same. This property can be understood intuitively by imagining that there are x beech trees, for example, at some time t. In the next generation, exactly 80% of the x trees become beeches, 10% become maples, and 10% become another species. We are left with the exact same x number of trees as before, just with different identities. The same logic applies to the y maple trees and z beech trees at time t. Multiplying by the transition matrix simply takes the $x + y + z = 100$ trees and divides them up differently among the three species groups, without gaining or losing any trees.

Although we have interpreted the vector $\boldsymbol{n}_t$ in this problem as a count of the trees that are members of each species, there is another equally valid interpretation of this calculation. The vector $\boldsymbol{n}_t$ can also be understood as giving the *probability*, in the frequentist sense, that any single randomly chosen tree in the patch is a member of each species. At a steady state, we thus can equivalently estimate that there is a 42% chance of a randomly selected tree being a beech, a 17% chance of a randomly selected tree being a maple, and a 41% chance of a randomly selected tree being another species. Transition matrices can also thus be thought of as modeling transitions in the probability that a random variable takes any particular value. In this problem, we can use the random variable S to represent the species identity of a single tree. At a steady state, the probability distribution for S would be given by

$$\Pr(S = B) = 0.42$$
$$\Pr(S = M) = 0.17$$
$$\Pr(S = O) = 0.41$$

The next chapter will expand on this probabilistic interpretation of transition matrices.

NEXT STEPS

From an ecological perspective, the simple transitions described by this model are an obvious simplification of the relatively complex process of succession. This model assumes, for example, that no disturbances occur to reset the forest to an earlier state and that the rates at which trees transition from one species group to another stay constant over time. This latter assumption could be very logically violated if, for example, the number of saplings of a species depends on the number of seeds of that species produced in an earlier year, which could be expected to change over time as the number of adult trees of each species changes.

In a formal sense, the type of transition matrix presented in this chapter is a **column stochastic matrix**, which is the type of matrix in which each column sums to 1. You will also find **row stochastic matrices** in the literature, in which each *row* of the matrix sums to 1 and the matrix is postmultiplied by a row vector, instead of a column vector. These two ways of setting up the model give identical answers; it's simply a matter of how to organize the numbers.

The data in this problem are simplified from a much larger analysis of succession presented by Horn (1975), who used a matrix model very similar to this one to examine a forest in Princeton, New Jersey.

14 **A Pair of Populations** Absorption

The park we visited in chapter 13 is one of two remaining seminatural habitat patches in its region. These two patches are inhabited by a threatened butterfly species. It has been observed that the species occasionally goes locally extinct in one patch, but the patch is then recolonized by butterflies from the other patch. For this particular species, the probability of local extinction in a given year appears to be around 13% in the smaller patch and 3% in the larger patch, and the probability of either empty patch being recolonized by butterflies from the other occupied patch in a given year is about 2%. Assuming that both patches are currently occupied, what is the probability that the threatened butterfly species will be permanently lost from both patches sometime in the next 50 years?

This problem is the first of several that we will encounter in this book that involve estimating the risk of extinction for a species. In this particular problem, our goal is to determine the probability that the butterfly species will be lost from both patches—in which case it presumably has no way of returning to this landscape—before a certain number of years have passed. Since the butterfly can potentially inhabit two different patches in this landscape, the butterfly will be lost entirely only if it becomes locally extinct in both patches simultaneously. This could occur either if both patches become empty simultaneously in the same year, or if one patch becomes empty and is not "rescued" by a colonist from the other patch before that patch also becomes empty. A population like this one that is spread across multiple patches in which there is a possibility of local extinction and recolonization is known as a **metapopulation**.

To begin, we can recognize that each of the two patches, which we will call A and B, can be found in two possible states, occupied or empty. There are thus four possible states in which we might find the metapopulation: both patches empty, A occupied and B empty, A empty and B occupied, and both patches occupied. For convenience, we'll label these four states as 0/0, 1/0, 0/1, and 1/1. In these labels, the first number indicates the state of patch A and the second the state of patch B, and a 1 indicates an occupied patch and a 0 indicates an empty patch. The entire metapopulation becomes extinct if it enters state 0/0. In the language of matrix models, a state that cannot be left once a system enters it is known as an **absorbing state**. Our goal in this problem is thus to estimate the probability that the metapopulation enters this single absorbing state 0/0 at any time before 50 years have passed.

We'll solve this problem using a matrix model similar to those that we have already encountered. Because we are interested in determining how a probability changes over time, in this case, the probability that the metapopulation is in any of the four possible states, we'll begin by building a transition matrix. Each row and column of the transition matrix A, from top to bottom or from left to right, will correspond to the states 0/0, 1/0, 0/1, and 1/1, in that order. To make this easier to visualize, we'll use a table to work out the numbers that we need to enter in this transition matrix.

	In 0/0 at t	In 1/0 at t	In 0/1 at t	In 1/1 at t
In 0/0 at $t+1$				
In 1/0 at $t+1$				
In 0/1 at $t+1$				
In 1/1 at $t+1$				

The interpretation of the numbers in each row and column of this table is similar to the interpretation of the rows and columns of the transition matrix in chapter 13. For example, the first row of the table gives the probability that the system enters into state 0/0 at time $t+1$ from each of the four possible states at time t, and the first column gives the probability that the system in state 0/0 at time t will transition into each of the four possible states at time $t+1$. As in all transition matrices, each column in the matrix must sum to 1, since the probabilities that a system in any current state will transition to each of the four possible states must add up to 1.

The next step is to fill in all of the 16 values in this table. This step will require some thinking. To help make it a bit simpler, we'll use e_A and e_B to represent the probabilities that the butterflies go locally extinct in each patch each year and c_A and c_B to represent the probabilities that each patch is colonized by immigrants from the other patch each year. Using A for the larger patch and B for the smaller one, we know from the problem that $e_A = 3\%$, $e_B = 13\%$, and $c_A = c_B = 2\%$. As we go through the next few steps, remember that a patch can become empty only if it is currently occupied, and that a patch can be colonized only if it is currently empty *and* the other patch is currently occupied.

Let's start with the metapopulation in state 0/0 at time t, which corresponds to the butterfly being extinct in both patches. The first column of the table gives the probabilities that a metapopulation in state 0/0 at time t will transition to each of the four states. As we know, the system cannot leave state 0/0 once it has entered that absorbing state, since there is no source of colonists that arrive in either patch. The probability that the metapopulation transitions from state 0/0 to state 0/0 is thus 1, and the probability that it transitions to any other state is 0. We can thus fill in the first column of the table as

	In 0/0 at t	In 1/0 at t	In 0/1 at t	In 1/1 at t
In 0/0 at $t+1$	1			
In 1/0 at $t+1$	0			
In 0/1 at $t+1$	0			
In 1/1 at $t+1$	0			

Next, let's consider state 1/1, in which both patches are occupied. Since both patches are already occupied, the only way for the system to transition to one of the other states is through one or both patches becoming empty. We can thus ignore any probabilities related to patch colonization for now. For a metapopulation in state 1/1 to transition to state 0/0, both patches must simultaneously become empty. If we assume that the probabilities of local extinction in the two patches are independent, this probability is given by $e_A e_B$. A system in state 1/1 can get to state 1/0 if patch A does not go extinct but patch B does, which has the probability $(1 - e_A)e_B$. By the same logic, a system in state 1/1 transitions to state 0/1 with probability $e_A(1 - e_B)$. Finally, for a system in state 1/1 to remain in state 1/1, both patches must avoid local extinction, which they do with probability $(1 - e_A)(1 - e_B)$. The fourth column of the table should contain these probabilities of state 1/1 transitioning to any of the four states.

	In 0/0 at t	In 1/0 at t	In 0/1 at t	In 1/1 at t
In 0/0 at $t+1$	1			$e_A e_B$
In 1/0 at $t+1$	0			$(1 - e_A)e_B$
In 0/1 at $t+1$	0			$e_A(1 - e_B)$
In 1/1 at $t+1$	0			$(1 - e_A)(1 - e_B)$

With a little bit of algebra, you can see that the four terms in the fourth column of the table add up to 1, as they must.

Next, we can consider the second column of our table, which gives the probabilities of the system transitioning from state 1/0 at time t to any of the four states at $t+1$. Since patch A is occupied in this state, it cannot be colonized. Instead, it can only become extinct or not become extinct. So we do not need to consider colonization probabilities for patch A. Conversely, patch B is empty in this case, and thus we need to consider only whether it is colonized or not. Moving in order through the states, we first calculate that transitioning from state 1/0 to state 0/0 requires patch A to become empty and patch B not to be colonized. The probability of this scenario is given by $e_A(1 - c_B)$. A metapopulation in state 1/0 will stay in state 1/0 if patch A does not go extinct and patch B is not colonized, the probability of which is given by $(1 - e_A)(1 - c_B)$. A metapopulation in state 1/0 will transition to state 0/1 if patch A goes extinct but patch B is colonized, which has the probability $e_A c_B$, and a metapopulation in state 1/0 will transition to state 1/1 if patch A does not go extinct and patch B is colonized, which has the probability $(1 - e_A)c_B$. You can once again show that these four probabilities sum to 1, and we can place them in our table to give

	In 0/0 at t	In 1/0 at t	In 0/1 at t	In 1/1 at t
In 0/0 at $t+1$	1	$e_A(1 - c_B)$		$e_A e_B$
In 1/0 at $t+1$	0	$(1 - e_A)(1 - c_B)$		$(1 - e_A)e_B$
In 0/1 at $t+1$	0	$e_A c_B$		$e_A(1 - e_B)$
In 1/1 at $t+1$	0	$(1 - e_A)c_B$		$(1 - e_A)(1 - e_B)$

Finally, we can use essentially the same logic as above to fill in the third and final column of our table, giving

	In 0/0 at t	In 1/0 at t	In 0/1 at t	In 1/1 at t
In 0/0 at $t+1$	1	$e_A(1 - c_B)$	$(1 - c_A)e_B$	$e_A e_B$
In 1/0 at $t+1$	0	$(1 - e_A)(1 - c_B)$	$c_A e_B$	$(1 - e_A)e_B$
In 0/1 at $t+1$	0	$e_A c_B$	$(1 - c_A)(1 - e_B)$	$e_A(1 - e_B)$
In 1/1 at $t+1$	0	$(1 - e_A)c_B$	$c_A(1 - e_B)$	$(1 - e_A)(1 - e_B)$

Using the values of e_A, e_B, c_A, and c_B from the problem, we can calculate the values of each of the cells in the table above. These values can then be placed in the same orientation in our transition matrix A, which becomes

$$A = \begin{pmatrix} 1 & 0.029 & 0.13 & 0.0039 \\ 0 & 0.95 & 0.0026 & 0.13 \\ 0 & 0.00060 & 0.85 & 0.026 \\ 0 & 0.019 & 0.017 & 0.84 \end{pmatrix}$$

Now that we have our transition matrix A, we need to determine the initial vector p_0, which contains the probabilities that our metapopulation is in any particular state at time $t = 0$. This vector p_0 has the same role as the vector n_0 in the last two chapters. The change in the variable name for the vector does not affect any calculations, but does help to remind us that we are now tracking changes in probabilities rather than in counts of individuals. We know from the problem that in the current year, the butterfly is found in both patches, and thus the metapopulation is in state 1/1. Our vector p_0 thus assigns a probability of 1 to state 1/1 and a probability of 0 to any other state:

$$p_0 = \begin{pmatrix} 0 \\ 0 \\ 0 \\ 1 \end{pmatrix}$$

Following the same procedures as in the last two chapters, we can now use our matrix to project the state of our metapopulation into the future using the equation $p_{t+1} = Ap_t$. Because the vector p_t contains the probabilities that the metapopulation is in any of the four states at any given time, the four numbers in this vector must always sum to 1 in all time periods t.

The instructions from chapter 12 can be used here in a new Google Sheet to project the probability that the metapopulation will have entered the absorbing state 0/0 in any year in the future. The instructions will simply need to be updated to allow for a 4 × 4 matrix and a vector with four possible states.

Projecting our metapopulation into the future, figure 14.1 shows the probability that the metapopulation will have entered the absorbing state 0/0 in or before all years up to $t = 50$. Notice that this curve can only increase and never decrease, since it is not possible for the

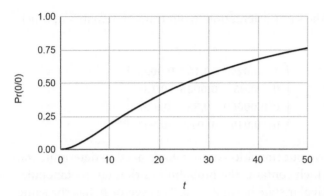

Figure 14.1. Probability that the metapopulation has entered the absorbing state 0/0 in or by a given year.

metapopulation to leave state 0/0 once it is entered. We find a 76% probability that the system will have entered state 0/0, in which the butterfly has become extinct in both patches, by year 50. Also interestingly, there is only a 3% chance that the system will be found in state 1/1, where it is found currently, in year 50. If we were to continue this calculation far into the future, eventually there would be a 100% chance that the butterfly would be extinct in both patches.

NEXT STEPS

The definition of an absorbing state is a state that a system, such as our metapopulation, cannot leave once it has been entered. As long as there is some probability that the system enters the absorbing state in each time step, the probability of the system being in the absorbing state will continue to increase until eventually, the system will be in the absorbing state with a probability of 1.

Given this inevitable result, you may wonder how any species is able to exist at all without eventually going extinct. At smaller spatial scales, the answer is that there is often a small but positive probability that a colonist arrives in the local population from elsewhere. This means that there is a small probability that the population is able to leave the state of complete local extinction, which would otherwise be absorbing. At larger spatial scales, the answer is that most species effectively have a large population spread across a large area, such that the probability of entering the absorbing state in any given year is extremely small. However, with that said, we should expect that given enough time, every species currently found on the planet will eventually go extinct, even leaving aside any impacts from humans.

This problem is based on a larger analysis of extinction risk across eight patches that was developed by Day and Possingham (1995). Like their analysis, ours assumes that extinctions in different patches are independent. In our simplified version of their analysis, we consider extinction and colonization simultaneously in a single matrix, rather than sequentially. Also note that Day and Possingham use a row stochastic matrix, instead of the column stochastic matrix that we use here.

15 **Fish Finders** Diffusion

> *Fish population surveys are currently being conducted in a nearby stream that is divided, for the purposes of the survey, into 1 m long segments. As part of these surveys, individual fish are being caught, tagged, and released back in the stream. The goal is to use the tags to study movement patterns in this fish population. If a fish has a 1% chance of moving to the next upstream segment and a 5% chance of moving to the next downstream segment each minute, how far from its initial location, on average, would you expect to find a fish one hour after it is released? How far might a fish have a realistic chance of moving in this period?*

Like the problem in the last chapter, this problem involves keeping track of a set of probabilities that change over time. In the last chapter, we built a matrix model to calculate the probabilities that a metapopulation would be in each of four possible states in or by any year t in the future. In this problem, our goal is to calculate the probabilities that a tagged fish will be found in any particular stream segment one hour after the fish has been released.

Compared with the problem in the last chapter, this problem has one important complication and one important simplification that we will need to incorporate into our matrix model. The complication is that there are many more than four possible stream segments, corresponding to states for our system, where a fish can be found after one hour. The problem doesn't state the overall length of the river, but a 1 km long river would have a thousand 1 m segments in which a fish could be found. This would seem to require using a vector with 1,000 numbers and a transition matrix with 1 million numbers in it!

To get started on this problem, however, we'll begin with a vector that tracks only the probabilities that a fish will be found in the segment in which it is released and the adjoining 10 segments upstream and 10 segments downstream. This gives us 21 possible segments to track. This is a larger set of states than we considered in the last few chapters, but still few enough that we can model them using the same basic approaches.

Even with this reduced scale, the transition matrix in this problem still has $21 \times 21 = 441$ numbers in it. Calculating all of these numbers using an approach similar to that in the last chapter would quickly become difficult. Fortunately, this problem allows a simplification that will make the task of filling in the transition matrix much easier. According to the problem, a fish has a probability $u = 0.01$ of moving upstream by one segment, and a probability $d = 0.05$ of moving downstream by one segment, in each minute. Importantly, the problem does not give a probability that a fish will move farther than one segment in a single minute. In our model, we will thus make the simplifying assumption that a fish cannot move more than one segment upstream or downstream in any minute. This assumption will greatly simplify the form of our transition matrix A and give it a very simple and recognizable structure.

Let's begin by considering our vector p_t, which will contain the probability that a fish is found at any possible segment in the stream at time t. As previously described, the vector p_t will have 21 rows, corresponding to the segment in which a fish is released along with 10 upstream and 10 downstream segments. We'll refer to the segment in which a fish is released as segment 0. The first segment upstream of the release location will be labeled 1, the next segment upstream will be labeled 2, and the 10th segment upstream will be labeled 10. Similarly, we'll label the first segment downstream of the release segment −1, the next segment −2, and the 10th segment −10. The 21 rows in the vector p_t will then correspond, in order from top to bottom, to the 21 stream segments running from 10 m upstream to 10 m downstream of the fish's release segment—that is, segments 10 through −10. The segment in which the fish is released, which is segment 0, corresponds to the 11th and center row of the vector p_t.

When we release the fish initially at $t = 0$, we know that it is found in segment 0. As discussed in the last chapter, this means that we can assign a probability of 1 to segment 0, right in the middle of the vector, and a probability of 0 to all other possible locations. Our initial vector p_0 will thus look like this:

$$p_0 = \begin{pmatrix} \vdots \\ 0 \\ 1 \\ 0 \\ \vdots \end{pmatrix}$$

The three vertical dots at the top and the bottom of this vector indicate that the vector continues in both directions. This shortcut saves us the trouble of having to write out the additional nine zeros above and nine zeros below the three numbers printed in the middle of the vector, which refer to segments 1, 0, and −1 from top to bottom.

Next, we can turn our attention to the transition matrix A, which will have 21 rows and columns. The columns of this matrix represent, from left to right, stream segments 10 through −10, and the rows of this matrix represent, from top to bottom, stream segments 10 through −10. As in the last chapter, the columns of the matrix indicate the stream segment where the fish is found at time t, and the rows indicate the stream segment where the fish will be found at time $t + 1$.

In the last chapter's problem, it was possible for the metapopulation to transition from any state to any other state in a single time step. In this problem, however, we will assume that a fish cannot move farther than one segment in a given time step. So, for example, if a fish is found in segment 9 at time t, there are only three possible segments where it can be found at time $t + 1$, which are 10, 9, and 8. These three possibilities reflect the probabilities that the fish will move upstream one segment, stay in the same segment, or move downstream one segment, in the next time step.

Using this basic idea, we can begin filling in the A matrix. To make this process easier to visualize, we'll begin by filling in the values in the top left-hand corner of the matrix. The first four rows and columns of this matrix are shown below.

$$A = \begin{pmatrix} ? & ? & ? & ? & \cdots \\ ? & ? & ? & ? & \cdots \\ ? & ? & ? & ? & \cdots \\ ? & ? & ? & ? & \cdots \\ \vdots & \vdots & \vdots & \vdots & \ddots \end{pmatrix}$$

The dots indicate that the matrix continues to the right and bottom of this corner, even though those values are not shown here. As in the A matrices in the previous chapters, each cell in this matrix represents the probability that the system transitions from one state to another during one time step, which here is equivalent to the probability that a fish physically moves from one stream segment to another.

Let's consider first the probability that a fish remains in the same segment from time t to time $t + 1$. Since the probability of moving upstream is $u = 0.01$, the probability of moving downstream is $d = 0.05$, and the probability of moving more than one segment upstream or downstream is zero, the probability of the fish remaining in the same segment is $1 - u - d = 0.94$. A fish thus has a 94% chance of remaining in the same stream segment from one minute to the next. As we've seen, the columns of the A matrix correspond to the location of the fish at time t, and the rows correspond to the location of the fish at time $t + 1$. The elements of the matrix along a diagonal line from the top left to the bottom right thus contain the probabilities that the fish remains in the same location at time $t + 1$. We can thus fill in the question marks along the diagonal of our A matrix as

$$A = \begin{pmatrix} 0.94 & ? & ? & ? & \cdots \\ ? & 0.94 & ? & ? & \cdots \\ ? & ? & 0.94 & ? & \cdots \\ ? & ? & ? & 0.94 & \cdots \\ \vdots & \vdots & \vdots & \vdots & \ddots \end{pmatrix}$$

Another way to see how this works is to imagine that we know at some time t that the fish is found in segment 9. The first four rows of our vector p_t would then be

$$p_t = \begin{pmatrix} 0 \\ 1 \\ 0 \\ 0 \\ \vdots \end{pmatrix}$$

The single value of 1 in vector p_t represents the probability of 1 that the fish is found in segment 9, which is the location represented by the second row from the top of this vector. When we perform the calculation $p_{t+1} = Ap_t$ to estimate the probability that the fish will be found in any given segment in the next time step, we will find that the vector p_{t+1} is

$$p_{t+1} = \begin{pmatrix} ? \\ 0.94 \\ ? \\ ? \\ \vdots \end{pmatrix}$$

which correctly assigns a 94% chance to the fish remaining in segment 9 at $t + 1$.

Let's continue to fill in more of the question marks in the A matrix, which will ultimately allow us fill in more of the question marks in the vector p_{t+1}. The labeling of the rows and columns of the transition matrix described above can help us see where the probabilities that a fish moves downstream should be entered. Based on this logic, the probability of the fish moving downstream, $d = 0.05$, should be entered in each column just below the probability of the fish remaining in the same segment. This updates our transition matrix to

$$A = \begin{pmatrix} 0.94 & ? & ? & ? & \cdots \\ 0.05 & 0.94 & ? & ? & \cdots \\ ? & 0.05 & 0.94 & ? & \cdots \\ ? & ? & 0.05 & 0.94 & \cdots \\ \vdots & \vdots & \vdots & \vdots & \ddots \end{pmatrix}$$

If we were to once again perform the calculation $\boldsymbol{p}_{t+1} = \boldsymbol{A}\boldsymbol{p}_t$ with this updated transition matrix, we would now find

$$\boldsymbol{p}_{t+1} = \begin{pmatrix} ? \\ 0.94 \\ 0.05 \\ ? \\ \vdots \end{pmatrix}$$

which reflects, as expected, the 5% chance that the fish moves downstream during the next one-minute period.

By similar logic, the probability that the fish moves upstream during the next time period, which is $u = 0.01$, should be entered in each column of the matrix just above the probability of the fish remaining in the same segment,

$$A = \begin{pmatrix} 0.94 & 0.01 & ? & ? & \cdots \\ 0.05 & 0.94 & 0.01 & ? & \cdots \\ ? & 0.05 & 0.94 & 0.01 & \cdots \\ ? & ? & 0.05 & 0.94 & \cdots \\ \vdots & \vdots & \vdots & \vdots & \ddots \end{pmatrix}$$

At time $t + 1$, we now have the vector

$$\boldsymbol{p}_{t+1} = \begin{pmatrix} 0.01 \\ 0.94 \\ 0.05 \\ ? \\ \vdots \end{pmatrix}$$

92 Modeling Multiple States

You might notice that we have encountered a complication in accounting for upstream movement, which occurs when the fish is found in segment 10. Since we decided above that we would set up our model to track only segments from 10 to −10, we are not recording any information about segment 11, the segment upstream of segment 10. We thus have no way of tracking the movement of a fish once it moves upstream from segment 10 and "falls off the map" at the boundary of our model. There are a few different ways that we could fix this problem. Here, we'll make the assumption that a fish in segment 10 can stay in that segment or move downstream, but cannot move upstream. We'll thus modify the number in the top left corner of our transition matrix to be the probability that the fish remains in segment 10 or attempts unsuccessfully to move upstream, so that the modified A matrix reads

$$A = \begin{pmatrix} 0.95 & 0.01 & ? & ? & \cdots \\ 0.05 & 0.94 & 0.01 & ? & \cdots \\ ? & 0.05 & 0.94 & 0.01 & \cdots \\ ? & ? & 0.05 & 0.94 & \cdots \\ \vdots & \vdots & \vdots & \vdots & \ddots \end{pmatrix}$$

As we will see below, when a fish begins in segment 0, there is only a very small chance that it will reach segment 10 or −10, the farthest segments that we are choosing to model, within one hour. The modification above thus has very little effect on our eventual solution to this problem.

Two final steps will allow us to complete matrix A. First, we can recognize that the values on and around the diagonal of the matrix continue down all the way to the bottom right corner. When we get to segment −10, we once again have the challenge of addressing the possibility that a fish might move downstream out of our system to segment −11. We can treat this case similarly to the case of a fish moving upstream from segment 10, as described above.

Second, we need to consider all of the values in the matrix that are not on or adjacent to the diagonal. These values have been denoted with ? symbols up to this point. Here, we will make use of our assumption that in a one-minute time period, a fish cannot move farther than an adjacent stream segment. This means that the probability of a fish moving beyond an adjacent stream segment in a single time step is zero. All of the remaining elements of the matrix shown with ? symbols, which represent the probabilities of a fish moving farther than one stream segment in one time step, can thus be replaced with zeros.

Fish Finders: Diffusion

This leads us finally to the complete form of our transition matrix:

$$A = \begin{pmatrix} 0.95 & 0.01 & 0 & 0 & \cdots & 0 & 0 & 0 & 0 \\ 0.05 & 0.94 & 0.01 & 0 & \cdots & 0 & 0 & 0 & 0 \\ 0 & 0.05 & 0.94 & 0.01 & \cdots & 0 & 0 & 0 & 0 \\ 0 & 0 & 0.05 & 0.94 & \cdots & 0 & 0 & 0 & 0 \\ \vdots & \vdots & \vdots & \vdots & \ddots & \vdots & \vdots & \vdots & \vdots \\ 0 & 0 & 0 & 0 & \cdots & 0.94 & 0.01 & 0 & 0 \\ 0 & 0 & 0 & 0 & \cdots & 0.05 & 0.94 & 0.01 & 0 \\ 0 & 0 & 0 & 0 & \cdots & 0 & 0.05 & 0.94 & 0.01 \\ 0 & 0 & 0 & 0 & \cdots & 0 & 0 & 0.05 & 0.99 \end{pmatrix}$$

We can now see all four corners of the matrix. The middle of the matrix continues with the values 0.05, 0.94, and 0.01 along the diagonal in between the top left and bottom right corners. Notice that we have replaced the value of 0.94 with 0.99 in the bottom right corner to correct for the boundary issue described above.

Returning finally to our initial question, recall that we have referred to the segment in which a fish is initially released as segment 0. This leads to an initial vector of

$$p_0 = \begin{pmatrix} \vdots \\ 0 \\ 1 \\ 0 \\ \vdots \end{pmatrix}$$

where the single probability of 1 is found in the 11th row of the vector, corresponding to the initial 100% probability that the fish is found in segment 0. As in the last few chapters, we can use the equation $p_{t+1} = Ap_t$ to update the vector for any time in the future, giving the probability that the fish is found in any particular stream segment at time t. Since we are interested in the fish's position after 60 minutes, we will perform this calculation 60 times and examine the vector p_{60}.

This problem is set up very similarly to those in previous chapters, except that the transition matrix and vector are larger. In a new Google Sheet, fill in the numbers 10 through −10 in Column A, beginning with Cell A2. Fill in the numbers 10 through −10 in Row 1, beginning with Cell B1. In Cells X1 and X2, enter d and 0.05, and in Cells Y1 and Y2, enter u and 0.01. Then, in each cell where 0.05 should appear, enter the formula =$X$2, and in each cell where 0.01 should appear, enter the formula =$Y$2. In each cell where 0.94 should appear, enter the formula =1-$X$2-$Y$2. You can use Copy and Paste to copy this formula from cell to cell to make things faster. In all cells where a zero should appear, enter 0. As described above, you'll need to correct the values in the upper left and bottom right corners of the matrix.

Starting in Cell A26, enter the numbers 10 through −10 down Column A, representing the values of the vector. Starting in Cell B25, enter the numbers 0 through 60 across Row 25, representing time steps. In Cells B26 through B46, enter the values of the vector p_0, which are 0 everywhere except for Cell C36, which has a value of 1. Use the approach from chapter 12 to project the vector to Column BJ.

The results of this calculation are shown in figure 15.1. This chart shows the probability distribution describing the probability that a fish is found in any of the 21 stream segments that we are tracking after one hour. Since the probability that a fish moves downstream is greater than the probability that it moves upstream in any time step, this distribution is shifted to the right of segment 0, the initial release point. The most likely segment for the fish's location is segment −2, corresponding to a 2 m shift downstream. Overall, there is a 98% chance that the fish will be found somewhere between segment 1 and segment −7 after one hour. The probability that the fish is found in segment 10 is nearly zero, and the probability that it is found in segment −10 is less than 0.05%. These results indicate that we would get effectively the same distribution if we had included distances farther from the release point in our model. You can confirm this result by expanding our model to track the location of the fish in stream segments at locations 15 to −15.

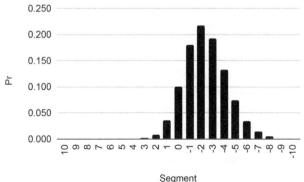

Figure 15.1. Probability that the fish is found in each of the 21 stream segments surrounding the release location after one hour.

The analysis that we have just performed is a discrete-time and discrete-space model of a process known as **diffusion with drift**. In models like the one in this chapter, diffusion describes how the probability that an object, such as a fish, is found in a particular location "spreads out" as time goes on. The process of diffusion leads eventually to a bell-shaped probability distribution like that in figure 15.1. In this problem, the mean of the distribution also shifts as t increases, which means that the average location of the fish shifts, or drifts, as time goes on. You can adjust the values for u and d in this problem to see how these two parameters affect the speed of both drift and diffusion. Notice that if the probabilities of a fish moving upstream and downstream in a time step are set to be equal, the distribution stays centered on segment 0, which is always the most likely location for the fish to be found. Also notice that as the values of u and d increase, or as the number of time steps increases, the fish is more likely to move farther from its initial release location.

NEXT STEPS

As we saw above, it turned out in this particular problem that the fish essentially never reached a location 10 m away from its initial release point within one hour. As a result, we did not have to think too hard about how to deal with the **boundary conditions** for our model, which determine how we treat the case in which the fish reaches the end of the area that we are modeling and could attempt to "leave" the set of locations we are tracking. In problems where boundaries are reached with some regularity, these are often modeled as either absorbing or reflecting. An absorbing boundary is akin to the absorbing state of two-patch extinction from chapter 14 in that it describes a state that the system cannot leave once it is entered. A reflecting boundary, in contrast, would assume that the fish cannot move downstream from segment −10 or upstream from segment 10 at all, as if a wall were placed at that location that reflects the fish back into the portion of the stream being surveyed. We set up our model above with a type of reflecting boundary.

In addition to modeling diffusion as a shift in location, a very similar matrix model can be used for modeling **birth-death chains**, which describe the stochastic increase or decrease in the size of a population by modeling individual births and deaths. A birth-death chain assumes that a population can increase or decrease by only one individual in a given time step, and as a result, the transition matrices for these models appear very similar to the one that we developed here.

This problem is based on an exercise presented by Otto and Day (2007), who use this example of fish movement in a stream to discuss continuous-space diffusion processes. See also Allen (2010) for details on birth-death chains and other models similar to the one in this chapter.

Part IV **Explaining Data**

16 Introducing Statistics

Part 4 of this book marks a bit of a transition from the previous three parts. In parts 1, 2, and 3, we solved many problems that required us to build models. Up to this point, each of the problems that we solved gave us information about the parameters of these models, such as growth rates, probabilities, or transition rates. In part 4, we'll begin to solve problems in which the model parameters are not given to us directly, but instead must be estimated from data.

Quantitative methods that combine data with models are broadly part of the field of **statistics**. Statistics, or even just ecological statistics, is an enormous area of study. There are not only entire books devoted to nothing but ecological statistics, but also entire journals, departments, and careers built entirely on this topic. Of course, in this book, we will only be able to scratch the surface of the breadth of statistical techniques used in ecology. In particular, we will discuss only the methods of frequentist statistics, which match the frequentist definition of probability that we used in part 2.

In frequentist statistics, probably the most foundational idea is the relationship between a **sample** and the **population** from which that sample is drawn. To understand this distinction, let's imagine that we would like to determine the average height of a tree in a particular forest. In this example, the set of all trees in the forest is considered our population. If we had unlimited time and energy, we could actually calculate the exact average height of a tree in this forest by measuring the height of every tree in the forest and averaging those numbers. It's more likely, however, that our time is limited, and that we will have to be content with measuring the heights of a small number of trees selected from this population. This set of trees whose heights we actually measure is called our sample, and the list of heights that we measure for this sample of trees makes up our data set.

Notice that the only thing that we've said so far about our sample is that it is a small selection of trees from the population. Importantly, we haven't said anything about how those trees were selected. You can imagine, for example, that if we chose the first trees that we encountered at the edge of the forest, we might accidentally select a sample of trees that are shorter and scrubbier than those in the interior of the forest. If we were to accidentally choose a sample of trees that were not representative of the entire population, our sample would be **biased**. This would lead our estimate of average tree height across the entire population to be incorrect. In the next four chapters, we'll avoid this issue of bias by assuming that the individuals in our samples were selected randomly from all individuals in the population. In our forest, we might do this by using a computer to generate pairs of random numbers that would correspond to the latitude and longitude of trees for us to measure. Avoiding bias in our sample is critical to frequentist statistics, since the ultimate goal of almost all frequentist statistics is to use information from a sample to estimate quantities in the entire population that we can't observe directly, such as average tree height across an entire forest.

Our basic approach to using information from our sample to learn about our population will be to build models, just as we have in many previous chapters in this book. In focusing on statistical analysis as a type of modeling, the next four chapters will probably appear very different from other presentations in introductory statistics books or classes that you may have seen. At introductory levels, it is much more common for statistics to be taught as a series of pre-defined calculations that can be applied to learn about data. This model-based approach, however, allows more advanced statistical methods to account for the real-world complexity of most field data collected by ecologists.

More so than in earlier parts of this book, the four chapters in part 4 are strongly sequential. I would encourage you to read all four chapters in order, as the concepts in each chapter follow directly from those discussed in the previous chapter. Ultimately, our goal will be to build a generalized linear model (GLM). Chapters 17 and 18 will focus on models that can be described by a single equation, which is a probability distribution describing the possible values of a single random variable. In chapters 19 and 20, we will then add a second nonrandom equation to these probability distributions. These two equations together will form a generalized linear model.

17 **Seedling Counts I** Maximum Likelihood

A friend who has an internship at a forest research station is attempting to determine what factors lead to the successful growth of a particular species of tree. She has counted the number of seedlings of this species in nine small plots, finding 5, 13, 0, 9, 8, 6, 6, 2, and 13 seedlings. Although she has sampled only a few plots, she would like to know the probability that a plot of the same size located somewhere in this forest contains no seedlings. What is your estimate of this probability?

At first glance, this problem appears very similar to the ones that we encountered in part 2 of this book. As in the problem involving the number of eggs per nest in chapter 7, we might begin by simply looking at the counts of seedlings in each of the nine plots and calculating probabilities directly from those. This approach leads to the probabilities shown in figure 17.1, where numbers of seedlings are on the x-axis and the probability of finding each number of seedlings in a plot is on the y-axis. We can see from this figure that there is a 1/9, or 11%, chance of finding a plot with zero seedlings in this set of nine plots.

However, if we think more carefully about the difference between a sample and a population, as described in the last chapter, we can see that this calculation does not really answer your friend's question. By asking for the probability that a plot somewhere in the forest contains zero seedlings, she is asking a question about the entire *population* of plots. Imagine, for example, drawing a very fine grid over the entire forest, resulting in a very large number of plots that could be selected for seedling counting. Your friend's question is about the probability of finding no seedlings in a plot chosen from this entire grid. The nine seedling counts your friend has collected, however, make up just one small sample of plots from this entire population. Our challenge here is thus to find a way to use information from our sample to estimate information about the entire population—in this case, the probability that a plot chosen from the entire forest contains zero seedlings. Although your friend doesn't state this in the problem, we'll assume that she means a *randomly* chosen plot within the forest, which will allow us to use the tools of frequentist statistics to solve this problem.

The challenge in extending information from our sample to the population is that the probabilities of observing particular seedling counts

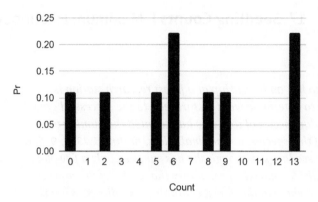

Figure 17.1. Probability that a plot in the sample of nine plots contains a particular number of seedlings.

in our sample are not necessarily a perfect match for the probabilities of observing particular seedling counts in the population. This is because there is some randomness in the number of seedlings observed in our relatively small sample of plots. It doesn't seem realistic, for example, that there could be zero plots in the entire forest with 3, 7, or 12 seedlings, even though we did not observe any plots with those exact counts in our small sample. It seems more likely that some counts were "missed" in our sample. If we had looked at a different, randomly chosen sample of nine plots, there's a very good chance that we would have observed a different set of seedling counts.

Let's start expanding our thinking from our sample to the entire population by considering the variable that we need to track to solve this problem. We can define the count of seedlings in a randomly chosen plot in this forest as a random variable Y. In any given plot, this seedling count variable will be equal to some number C with a probability $Pr(Y = C)$. As described in part 2, the set of probabilities $Pr(Y = C)$ for all possible values of C form a probability distribution for the variable Y that applies across the entire population of plots in the forest. The allowable number of seedlings in a plot ranges from zero to some very large number, the maximum number that could physically fit in a plot. This probability distribution will thus have support from zero to this very large number, which for convenience we will assume to be infinity. Once we have determined the shape of this probability distribution, we can use it to calculate $Pr(Y = 0)$, which will answer your friend's question.

To determine the shape of this population-level probability distribution, we will use the approach of **parametric statistics**, in which we assume that there is some particular equation that describes the shape of the probability distribution. When we have data that are counts, like counts of seedlings in a plot, it is very common to model this probability distribution as a Poisson distribution. We can then write a simple statistical model for the random variable Y as

$$Y \sim \text{Pois}(\mu)$$

The variable on the left-hand side of this equation, Y, is the number of seedlings found in a plot. Then, instead of an equals sign, the equation has a tilde (~) instead. This symbol is a shorthand way of indicating that the number of seedlings Y is a random variable that takes particular values with probabilities given by the right-hand side of the equation. The right-hand side, $\text{Pois}(\mu)$, indicates that we are using the Poisson distribution to give us the probabilities $\Pr(Y = C)$. The symbol inside the parentheses tells us that this distribution has one parameter, μ.

If you haven't encountered them before, probability distributions might seem like something strange or advanced. In reality, though, the names of different probability distributions are just shorthand ways of referring to particular equations that are used to calculate probabilities. In the case of the Poisson distribution, the equation used is

$$\text{Pois}(\mu) = \frac{e^{-\mu}\mu^c}{C!}$$

The right-hand side of this equation includes the number e, the parameter μ, and the count C for which we want to evaluate this probability. The number e in this equation is a special number, equal to about 2.71, that appears so often in mathematics that it's designated by a special letter, like π. The exclamation point (!) indicates the factorial of a whole number, which is the product of that number and all whole numbers less than it down to 1. For example, if our population of plots were described by a Poisson distribution with the parameter $\mu = 1$, we could calculate the probability that a randomly chosen plot had exactly 3 seedlings in it as $\Pr(Y = 3) = 2.71^{-1}1^3/3! = 2.71^{-1}1^3/(3 \times 2 \times 1) = 0.06$. Ultimately, our goal will be to use this Poisson distribution to calculate the probability of observing zero seedlings in a plot drawn from the entire population, which will answer your friend's question.

Before we move on, you may remember that the most basic rule of probability is that the probabilities of all possible values of a random variable must add up to 1. Any valid equation for a probability distribution, such as the one above, is carefully designed so that no matter what parameters are chosen, the sum of the distribution across its full support is always equal to 1. In the case of the Poisson distribution, this means that if we were to calculate Pr($Y = C$) for all values of C from zero to positive infinity, those numbers would always add up to 1.

Returning to the equation for the Poisson distribution, we can see that in order to use this equation to calculate the probability that a plot has C seedlings, we first need to know the parameter μ. Because we cannot survey every plot in the entire population, however, we will not be able to calculate the exact value of μ in practice. Instead, we will use the data from the sample of nine plots that your friend has surveyed to make our "best guess" for the value of μ, which we will call $\hat{\mu}$. The "hat" above the variable μ is used to indicate that it is our best estimate, based on our sample data, for the population parameter μ. The process of using sample data to determine unknown population-level parameters in a statistical model is called **fitting** the model.

To figure out the value of $\hat{\mu}$ for this model and data, we will use the **maximum likelihood** approach. To start, we will define **likelihood** as the probability of observing a data set given a particular model and set of parameters. Imagine, for example, that we counted the seedlings in one plot and found that there were 3. The likelihood of finding 3 seedlings in a plot for a population in which seedling counts follow a Poisson distribution with the parameter $\mu = 1$, for example, is equal to $e^{-1}1^3/3! = 0.06$.

We had no particular reason, however, to choose the parameter $\mu = 1$. The likelihood of observing 3 seedlings in a plot with this same model if $\mu = 2$, for example, would instead be $e^{-2}2^3/3! = 0.18$. Since the likelihood of a model with $\mu = 2$ is greater than that of a model with $\mu = 1$, we say that the model with $\mu = 2$ is more likely to describe the true population probability distribution. The principle of maximum likelihood states that our best guess for the value of the parameter μ in this Poisson model will be the value of μ that gives the largest possible likelihood. As described above, we will refer to this maximum likelihood estimate of μ as $\hat{\mu}$.

In the more complicated case presented in our problem, there is more than one data point available. The likelihood for all nine data points together is calculated as the product of the likelihood for all nine of the data points individually. Our goal is thus to find the value of μ that makes this likelihood for all nine data points as large as possible.

In practice, many likelihood values are so small that when we multiply them together, a computer will round down the product to zero. To help with this difficulty, it is common to take the natural logarithm of the likelihood for each data point and then add these together. This calculation gives the total **log likelihood** for the model and its parameters for our sample data. If, for example, L_1 and L_2 are the likelihoods calculated for two data points, then the likelihood of observing both data points is L_1L_2, and the log of this likelihood is $\log(L_1L_2) = \log(L_1) + \log(L_2)$. Maximizing the log likelihood of a model and its parameters will lead to the same best-fitting parameter estimates as maximizing the likelihood.

As an example, let's calculate the log likelihood of our nine observed seedling counts for a Poisson model with $\mu = 2$. The first observed count in this data set is 5. The likelihood of this observation is $e^{-2}2^5/5! = 0.0361$, and the natural log of this likelihood is -3.32. The second observed count in this data set is 13, which has a log likelihood of -15.54. The log likelihood of these two data points together is thus $-3.32 - 15.54 = -18.86$. Continuing for all nine data points, we find a total log likelihood of -62.17. Our task is then to find the particular value of the parameter μ that makes this number as large as possible.

> In the first row of a new Google Sheet, enter the text mu and sumlog(L) in Cells A1 and B1, respectively. Then, from Cell C1 to K1, enter the nine seedling counts given in the problem. In Cell A2, enter the value 1. In Cell A3, enter the formula =A2+0.25, and drag the corner of this cell downward to Row 58. This column now holds the values of μ for which we will calculate a likelihood.
>
> In Cell C2, enter the formula =LN(exp(-$A2)*$A2^C$1/fact(C$1)), which calculates the log likelihood for the first data point, a count of 5, for the parameter value $\mu = 1$. Drag the corner of this cell to the right, to Column K, to calculate the log likelihood for all nine data points for this value of μ. Then, in Cell B2, enter the formula =sum(C2:K2), which adds the log likelihoods for $\mu = 1$ across all nine data points. Highlight Cells B2 through K2 and drag downward to Row 58. Then review Column B to find the value of μ that maximizes the log likelihood.

Figure 17.2 shows that the value of $\hat{\mu}$ that maximizes the log likelihood for this Poisson model is around 7, which corresponds to a log likelihood of -29.5. Calculating the log likelihood at finer intervals would eventually show that the highest log likelihood is reached for a value of $\hat{\mu} = 6.9$. The particular value of a parameter that maximizes the likelihood of a model for a given data set is known as the **maximum likelihood estimate** of that parameter. Interestingly, this value of $\hat{\mu}$ is equal to the mean of the nine seedling counts observed by your friend. In our statistical model, the parameter μ also happens to represent the mean of the Poisson distribution. The maximum likelihood estimate of the mean of the Poisson distribution thus turns out to also be the mean of the sample data.

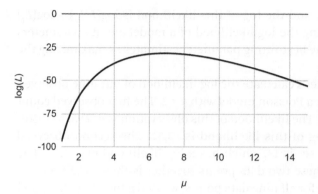

Figure 17.2. Log likelihood curve for the parameter μ. The maximum likelihood value of μ is near 7, corresponding to a log likelihood of approximately −29.5.

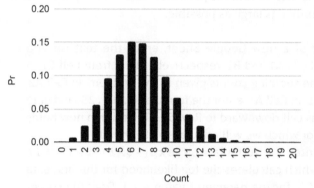

Figure 17.3. Modeled probability that a randomly chosen plot in the forest contains a particular number of seedlings. The probability of a plot containing zero seedlings is estimated to be 0.1%.

Now that we have calculated $\hat{\mu}$, we can calculate our best estimate of the probability that a randomly chosen plot will have exactly C seedlings, which is given by a Poisson distribution with the parameter μ = 6.9. These probabilities are shown in figure 17.3. Notice that according to our estimates, a plot is most likely to contain either 6 or 7 seedlings, a number similar to the mean of the sample data already collected. Based on these estimates, we can thus answer your friend's question: there is only a 0.1% chance of a plot chosen from the entire population having zero seedlings. This is much lower than the probability of 11% suggested by the sample data set itself.

To summarize our approach in this chapter, we began by proposing a simple model in which we assumed that the probability of observing a particular count of seedlings at the population level follows a Poisson distribution. We then used maximum likelihood, along with our nine sample data points, to find the maximum likelihood estimate of the parameter μ, known as $\hat{\mu}$, for this Poisson distribution. We then used the model with this best-fitting parameter value to calculate the probability of finding zero seedlings in a plot drawn from the population to answer your friend's question.

Although this approach may seem neat and tidy, we might wonder whether there are other models, particularly ones that assume different probability distributions, that we should have chosen. Perhaps some other model might be better than a Poisson distribution, and might even give a different answer to your friend's question. The next chapter will discuss how we can compare different models based on how well they appear to match our sample data.

NEXT STEPS

In this chapter, we have discussed maximizing the log of the likelihood of observing a data set as a way of finding the best-fitting parameters of a model. The idea of maximizing, or minimizing, the value of a function by changing the value of its inputs, known as **optimization**, is discussed in more detail in chapter 23. In this context, it's worth knowing that many optimization routines are specifically designed to minimize, rather than maximize, the value of a function. In practice, a computer will thus often find the maximum likelihood parameter values for a statistical model by instead minimizing the negative of the log likelihood function.

You might wonder about our assumption that there could be an infinite number of seedlings in a plot. Clearly, this is not literally true, as eventually there would be so many seedlings that another one could not physically fit within the plot boundaries. In this problem, however, we can presume that this maximum number of seedlings is very large and that the probability of observing such a large count is very small. In this scenario, the assumption of infinite support has little influence on our conclusions, since such large counts have almost no chance of being observed.

The data for this problem are based on redwood seedling counts collected by Strauss (1975) and analyzed in Diggle (2013).

18 **Seedling Counts II** Model Selection

> *As you read more papers that have analyzed data on seedling counts in plots, you come across a theory that suggests that these counts are best described by a geometric distribution, not a Poisson distribution. Would a model based on a geometric distribution give a better answer to your friend's question?*

The Poisson distribution, described in chapter 17, is widely used in ecology to model the probability of observing counts of organisms. However, it's not the only possible model. There are in fact ecological theories that predict such counts should follow different distributions, such as a geometric distribution, negative binomial distribution, or lognormal distribution. In this chapter, we'll compare models based on Poisson and geometric distributions to see which is more likely to provide an accurate answer to your friend's question.

We have already met the geometric distribution in chapter 10. In the context of counts of seedlings in plots, the geometric distribution states that the probability of observing C seedlings in a plot is given by

$$\Pr(Y = C) = \mathrm{Geom}(p) = (1-p)^C p$$

Like the Poisson distribution, which had one parameter, μ, the geometric distribution has one parameter, p.

We now have two potential models to describe the counts of seedlings in plots throughout the entire forest. In the last chapter, we proposed a model based on the Poisson distribution,

$$Y \sim \mathrm{Pois}(\mu)$$

and in this chapter we are proposing a model based on the geometric distribution,

$$Y \sim \mathrm{Geom}(p)$$

as an alternative.

Our goal now is to determine which of these two models is more likely to match the true probability distribution of seedling counts for the population of all plots in the forest. The fundamental challenge here, as described in the last chapter, is that we do not have access to information about counts of seedlings across the entire forest. Instead, we will need to find a way to use data from our smaller sample of plots in order to determine which of these two models is likely to be a better description of the entire population.

Comparing these two models will require two steps. First, we will need to find the best-fitting geometric model for our sample data, just as we did for the Poisson model in the last chapter. Second, we will need to compare the best-fitting Poisson model with the best-fitting geometric model. The task of comparing multiple models to determine the one that is most consistent with our observed sample data is known as **model selection**.

Our first step is thus to find the best-fitting geometric distribution for the nine seedling counts that we have observed. Since the geometric distribution has only one unknown parameter, p, we can use the approach from the last chapter to find the maximum likelihood estimate of p, which we will call $\hat{p}$. This approach finds a maximum log likelihood of -26.99, which occurs for $\hat{p} = 0.127$.

> The approach used in the last chapter to find the maximum likelihood estimate of μ for the Poisson distribution can also be used here to find the maximum likelihood estimate of p for the geometric distribution. Follow the same steps in a new Google Sheet, but modify the starting value of p in Cell A2 to be `0.001`, and the formula in Cell A3 to be `=A2+0.003`. This will allow us to examine smaller values of p at finer intervals. Then use the formula `=LN((1-$A2)^C$1*$A2)` in Cell C2. The rest of the steps remain the same.

Our best-fitting geometric distribution for seedling counts is shown as a dotted line in figure 18.1. This figure also shows the best-fitting Poisson distribution from the last chapter as a solid line, along with the histogram of the actual observed data. It turns out that the two models give fairly different predictions. The geometric distribution is more spread out than the Poisson distribution, with higher probabilities assigned to very low and very high numbers of seedlings. More importantly for your friend, the geometric model suggests that there is a 13% chance of a plot having zero seedlings, while the Poisson model suggests that there is only a 0.1% chance of a plot having zero seedlings. The geometric model thus suggests that the forest contains more than 100 times more empty plots than the Poisson model.

With these two models both fitted, we can now move on to deciding

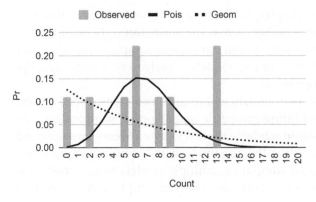

Figure 18.1. Observed probability of a plot in the sample containing a particular number of seedlings (gray bars), probability from the Poisson model that a randomly chosen plot in the forest contains a particular number of seedlings (solid line), and probability from the geometric model that a randomly chosen plot in the forest contains a particular number of seedlings (dotted line).

which of the two is more likely to describe counts of seedlings across the entire population of plots. Once again, since we have not actually counted the number of seedlings per plot in the entire population, we cannot know for sure which of the two models is a better description of the population. We can, however, use our sample data to help us guess which of the two models might provide a better match for the population-level distribution.

Up until this point, we have used the likelihood of a model only to determine the best-fitting parameters for that model. In doing this, we haven't actually paid any attention to the estimate of the likelihood itself, but only to the relative likelihoods of models with different parameters. You may have some intuition, however, that the likelihood estimate itself should provide us with information about how well the model matches the data in a more absolute sense. This is, in fact, a very reasonable interpretation. If a model has a high likelihood, it means that if that model were to be true, the data would have a relatively high probability of being observed.

Let's follow this logic and use it to compare the Poisson and geometric models. As we found in the last chapter, our best-fitting Poisson model for the sample data had the parameter $\hat{\mu} = 6.9$ and a log likelihood of -29.5. In this chapter, we found that our best-fitting geometric model had the parameter $\hat{p} = 0.127$ and a log likelihood of -26.99. The log likelihood for the geometric model is thus higher than the log likelihood for the Poisson model. We can interpret these results as stating that the nine seedling counts that we observed had a higher probability of being observed if the geometric model were the true description of the population of plots than if the Poisson model were the true description. This finding suggests that the geometric model is preferable and that we should use this model, rather than the Poisson model, to answer your friend's question.

Going beyond the relative rankings of the two models, there are also ways to use maximum likelihood to get a sense of how much better one model is than another. In ecology, one very common way of doing this is to use Akaike's information criterion, abbreviated AIC. The AIC for a model is calculated as

$$AIC = -2\ln(L) + 2K$$

On the right-hand side of this equation, $\ln(L)$ is the highest possible log likelihood for a model, which occurs when all parameters in the model are set to their maximum likelihood estimates. K is equal to the number of parameters in the model that had to be fitted using sample data.

In problems like this one, AIC values are used to provide a measure of the relative support for different models given some sample data that are available. Although the derivation and technical definition of the AIC are beyond the scope of this book, the basic idea is that the formula above estimates the information difference, up to an unknown constant, between the model being considered and an unknown true model that exactly describes the population. Even if the true model is not known, the difference between the AIC values of multiple models provides an estimate of how much closer one model is from the true, unknown model than the other.

Returning again to our problem, we can now calculate AIC values for the Poisson and geometric models given your friend's data. We have already found that the best-fitting Poisson model, which has one parameter μ that needed to be determined from the data, has a log likelihood of −29.5. Multiplying the log likelihood by −2 and adding 2 gives an AIC of about 61. The AIC for the geometric model, calculated similarly, is about 56.

Burnham and Anderson (2013) provide some often-cited rules for interpreting AIC values. In general, the model with the lowest AIC value has the strongest support. In our case, this is the geometric model, which, as we noticed earlier, had a higher log likelihood than the Poisson model. The difference in AIC between the geometric model and the Poisson model is about 5. Burnham and Anderson suggest that a model with a difference of 0–2 from the best-performing model still has substantial support, while differences of 4–7 indicate considerably less support, and differences greater than 10 indicate basically no support. In this case, with an AIC difference of 5 between the geometric and Poisson models, we can consider the geometric model to have substantially more support than the Poisson model. Your friend would thus be better served by presuming that the geometric model describes the probabilities of seedling counts across the entire forest.

NEXT STEPS

The distribution of counts of organisms across plots is a very widely studied probability distribution. In ecological statistics, it is very common to assume that counts follow a Poisson distribution. In macroecology, the Poisson distribution is generally thought to be a poor model for these counts, and alternative suggestions have included the geometric distribution (Harte 2011), negative binomial distribution (Green and Plotkin 2007), and lognormal distribution (May 1975). These alternative distributions generally lead to more plots with very small and very large numbers of individuals when compared with a Poisson distribution, which predicts a more even distribution of organisms across plots.

Akaike's information criterion is probably the most common quantitative metric used to compare the fits of multiple models. Burnham and Anderson (2013) provide an excellent and readable introduction that includes the technical details of the derivation and application of AIC. Burnham and Anderson also describe procedures for model averaging, in which AIC is not used to choose the "best" model, but instead to combine the predictions from multiple models based on their strength of support.

19 Flattened Frogs I Generalized Linear Models

Later in the year, you get an internship at a local government agency focused on wildlife protection. The agency is beginning a long-term study of the effects of roads on amphibian populations. They send you out to a highway with instructions to count the number of amphibian roadkills along two sections of the highway, one near a state park and one farther away. At four randomly chosen stops near the park, you find 11, 15, 13, and 3 roadkills, and at four randomly chosen stops far from the park, you find 8, 7, 9, and 7 roadkills. Is there a difference in the number of roadkills near and far from the park?

From an ecological perspective, counting road-killed amphibians may seem fairly different from counting tree seedlings in a forest. However, these two scenarios have much in common from a statistical perspective. Most importantly, in both cases, we have collected a small sample of data from a much larger population and want to use these data to answer a question about the population as a whole.

In the tree seedling example, we wanted to use data from our nine sampled plots to learn about the probability of finding certain counts of seedlings across all plots in the forest. In this roadkill example, we have data on roadkill counts from a sample of eight stops, some near the park and some far away, which were selected from a population of all possible stops along the highway. Our goal is to determine whether the number of roadkills across the population of all possible stops near the park is different from the number of roadkills across the population of all possible stops far from the park.

This extra complication of having two types of stops in our population, near and far, will eventually require us to design a more complex model than we used in the last chapter. Let's begin, however, with the simpler case of building a model to describe only the counts of roadkills in the section of highway near the park. In this simple case, all of the ideas from the last two chapters on tree seedlings apply to this problem, with the count of roadkills standing in for the count of seedlings and stops standing in for plots.

For the counts of roadkills at stops near the park, we might once again choose to construct a model based on a Poisson distribution. Calling the number of roadkills found at a stop Y, we once again have the model

$$Y \sim \text{Pois}(\mu)$$

where μ is an unknown parameter that equals the mean of the distribution. As before, this Poisson distribution can be used to calculate the probability of observing C roadkills at a stop, $\Pr(Y = C)$, given knowledge of the parameter μ.

Next, let's think about how we might extend this model to account for the fact that there are some stops located near the park and some stops located far from the park. We'll begin by imagining that every stop in our population, as well as every data point in our sample, is tagged with either an N or an F, which indicates whether that stop was located in the section of highway near or far from the park. We will then continue to assume that the probability of finding a certain number of roadkills at a stop is described by a Poisson distribution, but we will allow the parameter μ of the Poisson distribution to vary according to whether a stop is near or far from the park.

To do this, we'll use an expanded model that now consists of two equations that are used together, which are

$$Y_i \sim \text{Pois}(\mu_i)$$
$$\mu_i = \alpha + \beta X_i$$

Let's walk through the different parts of this model step by step. The first equation is almost identical to the one that we proposed earlier for the counts of roadkills in the section of highway near the park. Now, however, there's a subscript i attached to Y_i and μ_i. This i stands for whether the stop was located in the section of highway near or far from the park. We'll use a value of $i = 0$ to represent stops that are in the section of highway near the park and $i = 1$ to represent stops that are in the section of highway far from the park. When $i = 0$, we have $Y_0 \sim \text{Pois}(\mu_0)$, which says that the count of roadkills at a stop near the park follows a Poisson distribution with the parameter μ_0. When $i = 1$, we have $Y_1 \sim \text{Pois}(\mu_1)$, which says that the count of roadkills at a stop far from the park follows a Poisson distribution with the parameter μ_1. Thus, our model now allows the probabilities of finding certain numbers of roadkills at a stop to vary depending on whether that stop was near or far from the park.

The second equation in this new model tells us how to determine the values of μ_0 and μ_1 to use in the first equation. This equation states that the mean number of roadkills found near the park, μ_0, or far from the park, μ_1, is equal to a constant parameter α plus a second constant parameter β multiplied by a variable X_i. In this model, X_i is an indicator variable, or dummy variable, that indicates whether a particular stop was located near or far from the park. When $i = 0$ for a stop near the park, the indicator variable $X_0 = 0$, and when $i = 1$ for a stop far from the park, the indicator variable $X_1 = 1$.

Putting this all together, we see that our model can take two different forms, depending on whether we are considering the population of stops near or far from the park. For stops near the park, when $i = 0$ and $X_0 = 0$, the model suggests that the roadkill counts are described by

$$Y_0 \sim \text{Pois}(\mu_0)$$
$$\mu_0 = \alpha$$

For stops far from the park, when $i = 1$ and $X_1 = 1$, the model suggests that roadkill counts are described by

$$Y_1 \sim \text{Pois}(\mu_1)$$
$$\mu_1 = \alpha + \beta$$

For stops near the park, this model thus assumes that the number of roadkills is a random variable following a Poisson distribution whose mean μ_0 is equal to a constant α. For stops far from the park, this model assumes that the number of roadkills is a random variable following a Poisson distribution whose mean μ_1 is equal to the same constant α plus another constant β. We can thus see that α represents the mean number of roadkills at a stop near the park and $\alpha + \beta$ represents the mean number of roadkills at a stop far from the park.

The parameter β is thus the difference in the mean number of roadkills at stops near the park and at stops far from the park. This value is what we want to know in order to answer our question in this chapter. If β is greater than zero, it means there are more roadkills far from the park, and if it is less than zero, it means there are fewer roadkills far from the park.

It's worth knowing that the two parts of our model have specific names. The second equation in our model is called the **deterministic** part of the model. It can be useful to think of this part as giving the number of roadkills that are "supposed to occur" or "expected to occur" at a stop, depending on whether it is near or far from the park. The first equation in our model is called the **stochastic** part of the model. This part, which states that the actual observed number of roadkills is a random variable, introduces some amount of noise or randomness into the number of roadkills observed at any particular stop. The combination of a linear deterministic part and a Poisson stochastic part is known as a Poisson regression, which is one example of a **generalized linear model** (GLM).

Just as in the last two chapters, our next step after writing down the model is to fit the parameters of the model using maximum likelihood. In the full model that accounts for stops near and far from the highway, there are two unknown parameters, α and β. Notice that μ_0 and μ_1 are not actually parameters that have to be fitted, as they can be calculated from α and β. Notice also that X_i is not a parameter that has to be fitted, but rather part of our sample data. In this example, every stop in our sample has both a number of roadkills and a value of i associated with it, which tells us whether that stop was near or far from the park.

Unlike the models in the last two chapters, the model proposed here thus has two parameters rather than one. To fit our model to our sample data of eight observed roadkill counts, we need to find the values of these two parameters that together make the likelihood of our model as large as possible. We will use the notation $\hat{\alpha}$ and $\hat{\beta}$ to represent our best estimates of the unknown parameters α and β, respectively. Conceptually, finding the maximum likelihood estimates of these two parameters is very similar to the problem of fitting a one-parameter model. Let's consider, for example, proposed values of $\alpha = 10$ and $\beta = 2$. The first data point we have observed is a count of 11 roadkills at a stop near the park. For stops near the park, our model states that the number of roadkills $Y_0 \sim \text{Pois}(\mu_0 = \alpha = 10)$. The likelihood of observing 11 roadkills at a stop near the park given our proposed parameters is thus $e^{-10}10^{11}/11! = 0.11$. Another data point in our sample is a count of 7 roadkills at a stop far from the park. For stops far from the park, our model states that the number of roadkills $Y_1 \sim \text{Pois}(\mu_1 = \alpha + \beta = 12)$. The likelihood of observing 7 roadkills at a stop far from the park given our proposed parameters is thus $e^{-12}12^7/7! = 0.04$. Together, the likelihood of observing both 11 roadkills at a stop near the park and 7 roadkills at a stop far from the park, given a model in which $\alpha = 10$ and $\beta = 2$, is $0.11 \times 0.04 = 0.0044$. This is equivalent to a log likelihood of -5.43. We could continue with the other six data points from our sample to calculate the full likelihood for this model with these two parameters given our sample data.

Computationally, it can be a bit tricky to find the values of more than one parameter that together maximize the likelihood of a statistical model. This is an optimization problem, and it could be solved in a few ways. First, we might calculate the likelihood for a large number of combinations of values of both parameters and find the combination that gives the largest likelihood. This would be like repeating the approach of chapters 17 and 18, but in two dimensions, one for each parameter. Second, we might apply an optimization routine like the one we will describe in chapter 23. Third, we might try to find a trick that will allow us to take a shortcut to finding the optimal parameters. It turns out that this third approach can be used for this problem.

The trick that we can use to find the maximum likelihood estimates of α and β relies on the fact that the maximum likelihood estimate of μ in a Poisson distribution is equal to the mean of the sample data points. We first noticed this relationship in chapter 17, and we can use it again here in a slightly more complex form.

To start, let's consider the stops located near the park. The mean number of roadkills observed across the four stops near the park is 10.5. Using our "trick," this means that the maximum likelihood estimate of μ_0 is also 10.5. However, as we saw in the equations above, it turns out that $\mu_0 = \alpha$. We now know that $\hat{\alpha} = 10.5$ gives the maximum likelihood for our model. The next step is to find the best-fitting estimate of our other parameter, β. We can now consider the stops that are located far from the park. The mean number of roadkills across the four stops far from the park is 7.75. This means that the maximum likelihood estimate of μ_1 is also 7.75. Since $\mu_1 = \alpha + \beta$, and $\hat{\alpha} = 10.5$, we can figure out that $\hat{\beta} = -2.75$. The maximum likelihood parameters of our model are thus $\hat{\alpha} = 10.5$ and $\hat{\beta} = -2.75$, which together give a log likelihood of around −21.

At this point, we are ready to answer our question. Given the model that we have chosen to describe the entire population of possible stops on this highway, our data suggest that there are 2.75 fewer roadkills at stops far from the park than at stops near the park.

If we stop and consider this result for a moment, we might wonder why we bothered to go through building this model at all. In fact, a difference of 2.75 roadkills is exactly what we would have concluded if we had just compared the mean numbers of roadkills near and far from the park in our sample data. Building population-level models like the one in this chapter turns out be necessary, however, when we try to determine how much we should trust our estimate of this difference of 2.75 roadkills. Estimating our confidence in this estimate will be the topic of the next chapter.

NEXT STEPS

If you have some background in statistics, you might have noticed that the problem we solved here looked similar to some types of problems that appear in many introductory statistics books. The model that we built is actually very similar to the one that underlies a two-sample t-test. The main difference is that we have used a Poisson distribution for the stochastic portion of our model instead of the normal distribution that is used in a t-test. You may occasionally encounter a t-test being used to solve problems like this one, since the Poisson distribution begins to resemble a normal distribution once the mean parameter becomes reasonably large.

The Poisson regression presented in this chapter is an example of a generalized linear model, or GLM. More broadly, a GLM is a statistical model that has a linear deterministic component, a stochastic component that comes from the exponential family of probability distributions, and a link function that relates the deterministic component to the mean of that probability distribution. Many commonly used statistical models, including traditional linear regression (normal distribution and identity link) and logistic regression (Bernoulli distribution and logit link), can be interpreted as types of GLMs. It's worth noting that we constructed a Poisson regression in this chapter using an identity link function, whereas the more common form of a Poisson regression uses a logarithmic link function. Zuur et al. (2009) provide a clear and readable introduction to this class of models.

The data in this problem are drawn from a study of amphibian road-kills in Portugal presented by Zuur et al. (2009). There are many more data points and variables in the original data set, and Zuur et al. analyze these data with a much more complex model than the one that we built here.

20 **Flattened Frogs II** Hypothesis Testing

You report back to your supervisor that you believe that there are about 2.75 fewer roadkills at stops far from the park than at stops near the park. Your supervisor is not convinced and suggests that since you sampled a total of only eight stops, a small difference like this might be due to random chance and not due to any "real" effect of the park. On the basis of the data that you've collected, can you reject the idea that this difference is due to random chance?

This chapter gets to the heart of frequentist statistical analysis, which is very often centered on the concept of **hypothesis testing**. The purpose of hypothesis testing is to answer questions such as the one posed in this chapter's problem, which is how we can tell whether an effect that we observe in our sample data is "real" or whether it could have arisen due to random chance.

To address your supervisor's question, we'll return to the model that we developed in the last chapter. There, we assumed that the number of roadkills observed at a stop would be a random variable following a Poisson distribution, and that the mean of this Poisson distribution would be different for stops near the park and stops far from the park by an amount β. This led to the model

$$Y_i \sim \text{Pois}(\mu_i)$$
$$\mu_i = \alpha + \beta X_i$$

Once again, Y_i is the number of roadkills observed at a particular stop, and X_i is an indicator variable that has a value of 0 for stops near the park and a value of 1 for stops far from the park.

This model has two parameters, α and β, whose values we can estimate by fitting the model to our data using maximum likelihood. Our estimate of the difference in roadkills between stops near and far from the park is equal to the value of the parameter $\hat{\beta}$. In chapter 19, we found maximum likelihood estimates of $\hat{\alpha} = 10.5$ and $\hat{\beta} = -2.75$ for our data set. Our best estimate is thus that there are 2.75 fewer roadkills per stop far from the park than near the park.

When your supervisor questions whether the difference in the number of roadkills near and far from the park might be due to "random chance," she is effectively questioning how sure you are about your estimate of β. Remember that the true value of β refers to the difference in the mean number of roadkills at stops near and far from the park across the entire population of stops. In principle, this number could be calculated exactly by measuring the number of roadkills at every possible stop near and far from the park and comparing those numbers. As always, however, we do not have access to data from every stop near and far from the park. Thus, we can never be entirely certain about the true value of β because it describes the entire population of stops, which we were not able to observe. Our best estimate of β, which is $\hat{\beta} = -2.75$, comes just from the eight stops in our sample.

Although she hasn't stated it explicitly, your supervisor is actually asking a more specific question than how certain you are about your estimate of β. She's asking, in particular, whether you can rule out the possibility that in reality, $\beta = 0$, and there is thus no difference at the population level between the mean counts of roadkills at stops near and far from the park. Let's imagine for a moment that the true value of β is zero. Because the number of roadkills observed at a stop is a random variable following a Poisson distribution, any particular stop could have a roadkill count higher or lower than the expected average. With only four stops sampled near the park and four stops sampled far from the park, it's possible that just by chance, the four stops near the park happened to have counts that were higher than average, and the four stops far from the park happened to have counts that were lower than average. If this occurred, it might appear to us that the stops near the park had more roadkills, but this would be due to random chance, not to any actual difference at the population level.

There are many different ways that we could conduct a hypothesis test to determine whether we can confidently reject the idea that $\beta = 0$ and that the mean numbers of roadkills at stops near and far from the park are the same. Here, we'll use a test based on **parametric bootstrapping**, which will make all of the various steps needed to conduct a hypothesis test very clear. Often, the details of how statistical tests are conducted are hidden by the software used to conduct those tests. Our parametric bootstrap test, in contrast, will allow us to see every step explicitly. Fortunately, all of the concepts below apply to many other hypothesis tests that you might want to perform using other data sets and models.

The basic framework of hypothesis testing has five steps. First, we propose a statistical model to describe our system, which we have already done. Second, we propose a **null hypothesis**, which often expresses the idea that the effect we are interested in does not actually exist. Third, we choose a test statistic to evaluate our null hypothesis and determine its null distribution. Fourth, we determine the *P*-value for our test, which tells us how certain we can be that an effect that we see in our sample data is not due to random chance. Fifth, we interpret the *P*-value to accept or reject our null hypothesis and draw the appropriate scientific conclusion. Although this might seem like a lot of new ideas all at once, we'll walk through this whole process step by step.

The first step in a hypothesis test is to choose a statistical model, with both a deterministic and a stochastic component, that we believe describes the population from which we drew our sample. In this example, we have already developed a model with a linear deterministic component and a Poisson stochastic component. Although we can never know this for sure, we assume that this model provides a reasonably good description of how the number of roadkills at each stop is actually related to distance from the park at the population level.

The second step in a hypothesis test is to propose a null hypothesis that we wish to test. Your supervisor has suggested what is probably the most common type of null hypothesis, which is that the true value of some parameter in the model is zero. In this particular case, the null hypothesis proposed by your supervisor is that $\beta = 0$, which would indicate that there is no difference in the mean number of roadkills at stops near and far from the park at the population level. The alternative to this null hypothesis is that the value of β is, in fact, not zero. The idea is that if we have enough of the right kind of evidence in our data, we will be able to reject the null hypothesis and conclude that the mean number of roadkills at a stop does vary with distance from the park.

The third step is to choose a test statistic that we will use to evaluate our null hypothesis by calculating its null distribution. The null distribution is at the core of hypothesis testing and is the most important concept in this chapter. To begin, let's imagine a world where our null hypothesis is true and the parameter β is, in fact, zero. In this imaginary world, our model says that the number of roadkills at any of the eight stops, Y, will be a random number drawn from a Poisson distribution with a mean of μ, which is also equal to α in the deterministic part of our model. Thinking back to the simple model that we built and fit in chapter 17, we can calculate that our best-fitting estimate $\hat{\mu}$ for this simple model will be 9.125, the mean number of roadkills across the eight stops in our data set.

We can use this result to conduct a parametric bootstrap test of our null hypothesis. Let's imagine that we live in a world where our null hypothesis is true and that we conduct our survey one time. Each of our eight roadkill counts will be a random number drawn from a Poisson distribution with a mean of 9.125. By drawing eight random numbers from this distribution, I ended up with an imaginary survey in which I observed 11, 13, 5, and 6 roadkills at the stops near the park and 8, 7, 9, and 7 roadkills at stops far from the park.

In a new Google Sheet, in Row 1, enter the letter N in the first four columns and the letter F in the next four columns. The four columns with an N will contain roadkill counts for the four stops near the park, and the four columns with an F will contain roadkill counts for the four stops far from the park. In Row 2, enter the numbers 11, 15, 13, 3, 8, 7, 9, 7 in the first eight columns, which represent the real survey data we collected. To conduct one imaginary simulated survey, enter the formula =max(0,ROUND(NORMINV(RAND(),9.125,SQRT(9.125)))) in Cell A3, then highlight and drag this cell over to Cell H3. Since there is no easy way to draw a random number from a Poisson distribution in Google Sheets, this formula instead draws a random number from a normal distribution, rounds this random number to the nearest integer, and ensures that it is greater than or equal to zero. This process gives a close approximation to drawing a random number from a Poisson distribution with a mean of 9.125.

The critical idea behind any hypothesis test can be stated as follows: *if the data that we actually observed are very inconsistent with the data that we would observe if the null hypothesis was true, then we are justified in concluding that the null hypothesis is not true.* To perform our hypothesis test, we conduct a very, very large number of imaginary surveys in our imaginary world where the null hypothesis is true. We then compare our data from our one real survey with the "data" from these imaginary surveys to see if our real data are reasonably likely to be observed if the null hypothesis is true. If our data take a form that is very unlikely to occur in a world where the null hypothesis is true, then we conclude that we are justified in believing that we do *not* live in a world where the null hypothesis is true.

Putting this idea into practice requires a few steps. First, we need to choose a way to summarize the results from our one real survey and from all of our imaginary surveys with a single number. This number, known as the **test statistic**, provides us with a way of comparing our real data with the imaginary data sets. For our parametric bootstrap test, we can use the maximum likelihood estimate $\hat{\beta}$ as our test statistic. For our real data set, we have already determined that $\hat{\beta} = -2.75$, and this will be the value of the test statistic for our one real survey.

We noted above that one way of stating our null hypothesis is that $\beta = 0$ at the population level. Thus, in an imaginary world where the null hypothesis was true, we would generally expect our surveys to estimate that $\hat{\beta}$ is equal to or close to zero. However, when we use our "trick" from the last chapter to find $\hat{\beta}$ for the one simulated survey described above, we instead find $\hat{\beta} = -1.25$. In fact, if we create many imaginary surveys, we will find that $\hat{\beta}$ for any given survey may actually be quite far from zero. This phenomenon highlights exactly the issue that concerns your supervisor in this problem. Even if there is actually no difference in the mean number of roadkills at stops near and far from the park at the population level, so that $\beta = 0$, randomness can sometimes give us a data set in which it appears that a difference exists when we calculate a value of $\hat{\beta}$ that is not zero.

> Enter alphahat in Cell I1 and betahat in Cell J1. In Cells I2 and J2, enter the formulas =average(A2:D2) and =average(E2:H2)-I2, which apply the "trick" from the last chapter to quickly calculate the maximum likelihood estimates $\hat{\alpha}$ and $\hat{\beta}$ for our one real survey. Next, highlight Cells I2 and J2 and drag these down to Row 3 to calculate $\hat{\alpha}$ and $\hat{\beta}$ for our first imaginary survey. Your value of $\hat{\beta}$ will probably be different from the value of -1.25 that I reported above, since your randomly generated counts of roadkills at each stop will not match the ones above.

Let's pursue this idea a bit further. Imagine now repeating our imaginary survey not once, as we've done so far, but many times. When we do this, we can construct a distribution of the values of the test statistic that we found across all of our imaginary surveys in our imaginary world where the null hypothesis is true. This distribution is known as the **null distribution**. The null distribution for 1,000 calculations of our test statistic $\hat{\beta}$ from 1,000 imaginary surveys is shown in figure 20.1. The most important lesson that we learn from figure 20.1 is that even in a world where our null hypothesis is true, we sometimes get values of $\hat{\beta}$ that are far from zero.

> To simulate the null distribution, highlight all of the cells in Rows 1 and 2 and make the text in those cells **bold** to keep them clearly separate from the imaginary data that will be added below. Highlight

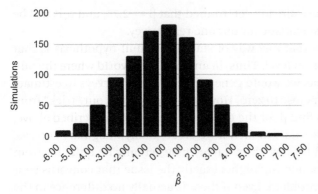

Figure 20.1. Simulated null distribution of the test statistic $\hat{\beta}$ under the null hypothesis that there is no difference at the population level in the mean number of roadkills at stops near and far from the park.

Cells A3 through J3 and drag these all the way down to the bottom of the Google Sheet until you get to Row 1,002, for a total of 1,000 simulations. When you get near the bottom, you may need to add more rows to the sheet.

Column J from Row 3 through Row 1,002 now contains values of $\hat{\beta}$ for 1,000 imaginary surveys in a world where the null hypothesis is true. Highlight these cells and go to Insert → Chart. In the Chart editor to the right, choose Histogram from the Chart type box to visualize the null distribution. Go to the Customize panel, under Histogram, and change the Bucket size to 1 to replicate the appearance of figure 20.1.

With our null distribution in hand, we are now ready to move on to the fourth step of hypothesis testing, which is calculating a **P-value** for our test. P-values are very widely quoted when explaining the results of hypothesis tests, but they are also unfortunately widely misunderstood, or at least partially misunderstood. The definition of a P-value is *the probability that we would observe a test statistic at least as extreme as our real, observed test statistic in a world where our null hypothesis was true*. If the test statistic that we actually observe is not likely to be observed by chance in a world where the null hypothesis is true, then we will find a very small P-value. If, on the other hand, the test statistic that we actually observe is relatively likely to be observed by chance in a world where the null hypothesis is true, then the P-value will be large.

Calculating the P-value in this case requires determining what fraction of our imaginary surveys have a test statistic at least as far from zero as our real, observed test statistic of $\hat{\beta} = -2.75$. When I constructed the null distribution shown in figure 20.1, I found that 216 of the 1,000 values of $\hat{\beta}$ from my 1,000 imaginary data sets were greater than or equal to 2.75 or less than or equal to −2.75. This gives us a *P*-value of 216/1,000 = 0.216 for our test.

To count the number of simulated test statistics farther from zero than our one real test statistic, go to Cell K2 and enter the formula =countif(J3:J1003,"<=-2.75"), which will count the number of estimates of β from Cells J3 through I1002 that have a value less than or equal to −2.75. In Cell L2, enter the formula =countif(J3:J1003,">=2.75") to calculate the number of values of $\hat{\beta}$ that are greater than or equal to 2.75. The sum of these two numbers gives the number of $\hat{\beta}$s from our simulated surveys that are at least as extreme—that is, at least as far from zero—as our one real $\hat{\beta}$ from our actual data set. This number will not be exactly 216, as described above, since your imaginary data sets will be different from mine.

Finally, we can proceed to the fifth and final step of hypothesis testing, which is drawing the appropriate conclusion based on our *P*-value. As we've seen above, the *P*-value describes the probability that if we lived in a world where the null hypothesis was true, we would observe a test statistic at least as extreme, or as far from zero, as the one that we observed in our actual survey. In this example, we found *P* = 0.216, indicating an approximately 22% chance of obtaining a test statistic farther from zero than the one from our real data set if the null hypothesis was true.

The appropriate interpretation of *P*-values remains a topic that is debated by both statisticians and ecologists. By scientific convention dating back to the early days of statistics, however, a *P*-value of 0.05 is often taken to provide a threshold for concluding that an observed test statistic is too extreme to have occurred by chance if the null hypothesis was true. The Greek letter α, not to be confused with the parameter α from our model, is commonly used to refer to this threshold. Thus, when we see a *P*-value less than α = 0.05, we often conclude that we have sufficient evidence from our data to reject the null hypothesis, which in this example is the hypothesis that there is no difference in the number of roadkills between stops near and far from the park. Since our *P*-value is greater than 0.05, we conclude that we do *not* have sufficient evidence to reject the null hypothesis. Thus, at the end of this exercise, you are forced to agree with your supervisor that in fact, the apparent difference in roadkill numbers near and far from the park in your survey might indeed be due to chance and not indicative of any true underlying difference at the population level.

NEXT STEPS

The idea behind hypothesis testing is explained and illustrated clearly in Whitlock and Schluter (2019), which is the first resource that I would recommend for expanding beyond the introduction above. Whitlock and Schluter do not cover generalized linear models, however, and I would recommend Zuur et al. (2009) as a second book that includes more advanced material.

To understand a bit more about hypothesis testing, it's important to recognize that there are two kinds of errors that we might make when performing and interpreting a hypothesis test. The first kind of error, known as type I, is rejecting the null hypothesis even though the null hypothesis is true. In other words, we might conclude that distance from the park has an effect on roadkill numbers even though, at the population level, it does not. The probability of making a type I error is always equal to α. The second kind of error, known as a type II error, is not rejecting the null hypothesis even though the null hypothesis is false. In our example, this would occur if we concluded that distance from the park did not matter even if, at the population level, there was in fact a difference between roadkill counts near and far from the park. Given that we did not reject our null hypothesis in this example, we could not have made a type I error, but it is possible that we may have made a type II error. This kind of error is particularly likely in cases, like this one, when data sets are very small.

As mentioned above, there is often more than one approach that can be reasonably used to test a null hypothesis for any given statistical model and scientific question. Our parametric bootstrap method above used $\hat{\beta}$ as our test statistic and drew random numbers in order to simulate our null distribution. This is a perfectly valid approach to hypothesis testing, and in fact is one of the preferred approaches for hypothesis testing involving complex statistical models. However, for simpler models, you will more often see hypothesis tests conducted using test statistics for which the null distribution can be expressed using a known equation. Some particularly common tests include the *t*-test, Wald test, and likelihood ratio test. These tests are explained in more detail in Whitlock and Schluter (2019) and Zuur et al. (2009).

Part V Expanding the Toolbox

21 Other Techniques

In this final part of the book, we will work through a series of problems that are united mainly by their inability to fit neatly into any of the other parts. In chapters 22 and 25, we will introduce two new and important ways of answering ecological questions, graphical methods and computational methods, that we have not encountered before. In chapter 23, we will dig deeper into the idea of optimization, which is an important topic in its own right and is also needed for solving many problems of the type posed in part 4. In chapter 24, we'll combine ideas from parts 1 and 2 of the book to conduct a simple stochastic simulation. Unlike parts 1–4 of this book, there is no particular background that you'll need before continuing on to chapters 22–25.

As we approach the end of this book, I hope that you are beginning to get a better sense for how various quantitative methods can be used to answer questions in ecology and conservation, and that you are feeling more confident in your ability to understand and use these methods. Most importantly, I hope that these problems have inspired you to continue your reading and studies beyond the specific topics that we have covered. We can call this book a success if you are able to use what you have learned here as a springboard into applying new and creative quantitative thinking to your own ecological problems, whatever those may be.

22 Bird Islands Graphical Thinking

> *Two offshore islands, one close to the mainland and one farther away, are both home to several species of birds. Historically, both islands have hosted the same average number of bird species, although the particular species on each island change over time as new colonists arrive from the mainland and species present on the island go locally extinct. Which island would you expect to have more available habitat for birds? Which island is likely to experience more frequent turnover of the particular species present there?*

This problem might initially read as something of a riddle. Unlike the other problems posed in this book, this problem does not contain a single number describing the bird species or the islands. Instead of approaching this problem with equations and calculations, we will thus approach it with a series of plots that will, perhaps surprisingly, allow us to answer the two questions above in a very general sense.

To begin, we need a framework that allows us to think about how the number of species on an island is determined. We will rely on the well-known theory of **island biogeography** to guide our thinking. Island biogeography proposes that the number of species found in an isolated habitat arises from a balance between colonization and local extinction. In short, species that are not present on an island may arrive through dispersal from the mainland, increasing the number of species on the island. However, at the same time, species already present on the island may go locally extinct, either through competition or through random events. When the rates at which species arrive and go extinct are equal, the number of species present on the island will reach a stable, steady-state value. The average number of species on the island will then stay the same over time, even as the actual identities of the species may change.

Let's attempt to put the above logic into the form of a plot. So far in this book, we have made many plots showing the results of our analyses. These plots have all had an explanatory variable on the x-axis, which was often time, and a response variable on the y-axis, which was often population size. We have used these plots to visualize how the variable on the y-axis changes as the variable on the x-axis changes. For this problem, we are going to continue making plots of a response variable on the y-axis against an explanatory variable on the x-axis, but the identities of these variables are going to change in a way that might initially seem a bit confusing.

Let's start by returning to the first equation that we saw in this book, which was the exponential growth model from chapter 2. There, we saw that exponential growth leads to a pattern of population change over time that looks like figure 22.1. As usual, in this figure, time is on the x-axis and population size is on the y-axis.

Let's now create a plot that shows this equation a bit differently. The difference equation for exponential growth can be written as

$$N_{t+1} = N_t + rN_t$$

where r is the per capita population growth rate. Instead of projecting how a population following this model might change over time, let's instead make a plot that shows how the amount by which the population changes in the next time step depends on the current population size. To do this, we subtract N_t from both sides of the equation above, giving

$$\Delta N_t = N_{t+1} - N_t = rN_t$$

The triangular symbol in the equation above is read as "delta." The variable ΔN_t represents the change in the population size in the next time step, given that the population currently has the size N_t. Figure 22.2 plots ΔN_t, the change in population size in the next time step, on the y-axis and N_t, the current population size, on the x-axis.

The key to understanding this new plot is to recognize that the line does *not* show how population size changes over time. Whereas the variable t appeared on the x-axis in figure 22.1, it does not appear on either

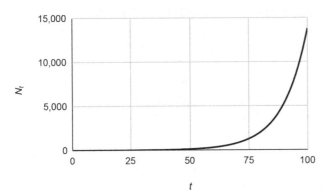

Figure 22.1. Exponential growth in a hypothetical population over time.

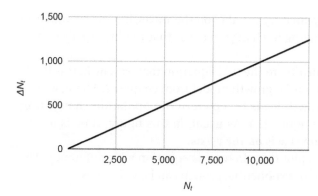

Figure 22.2. Change in population size in the next time step for a given population size in the current time step, based on the same exponential growth model as figure 22.1.

axis in figure 22.2. Instead, it's best to read this plot by imagining that we have encountered a new population with some initial size N_t that is between 0 and 12,500. Imagine, for example, that we have encountered a population that has 5,000 individuals. Looking along the x-axis in figure 22.2, we find the population size $N_t = 5,000$, and read along the y-axis that the corresponding ΔN_t will be 500. This means that 500 individuals will be added to the population in the next time step, bringing the total number of individuals in the next time step to $N_{t+1} = 5,500$. If we wanted to, we could then find the value of N_{t+2} by looking along the x-axis for 5,500 and finding the associated ΔN, and so on.

With this new plotting framework in mind, we can now return to the theory of island biogeography and our questions about the bird communities on the two islands. As described earlier, the theory of island biogeography proposes that the number of species in an isolated habitat, like an island, will increase due to colonization and decrease due to extinction. Our goal will be to create a plot, similar to figure 22.2, that shows how these two processes combine to determine the number of species on an island. We will use the variable S_t to track the number of species present at time t on our x-axis and the variable ΔS_t to track the change in the number of species during the next time step on our y-axis.

Let's begin by creating a plot that shows how extinction might change the number of species on an island in the next time step, given the number that are currently on the island. When there are zero species on an island, there cannot be any extinctions in the next time step, since there are no species to go extinct. When S_t is zero, ΔS_t due to extinction must also be zero. As S_t gets larger, it's reasonable to expect that the number of extinctions is also likely to increase as more species are forced to share the fixed resources of the island and more species are potentially subject to random disasters. We don't know exactly how ΔS_t will increase with S_t,

134 **Expanding the Toolbox**

but for simplicity we'll show this increase as a straight line. Figure 22.3 shows our proposed extinction curve.

We can follow similar logic to determine how colonization from the mainland will change the number of species on an island. On non-evolutionary time scales, the number of species found on the mainland will be some fixed number. Once the number of species on the island reaches this maximum, the number of species that can colonize the island will be zero. Conversely, when there are no species present on the island, the number of species colonizing the island in the next time step should be greater than zero. This means that the colonization curve should decrease from some larger number when there are few species until it reaches zero when S_t equals the number of species on the mainland. We can thus draw a proposed colonization curve as in figure 22.4.

These colonization and extinction curves show how the number of species on an island will increase or decrease in the next time step as new species colonize the island and existing species become locally extinct. Because both curves have the same x-axis and y-axis, we can combine the two curves into one plot, as shown in figure 22.5. The line that

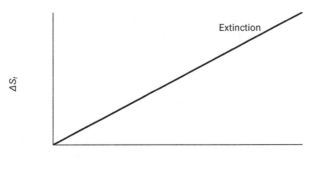

Figure 22.3. Hypothetical change in number of species due to extinction in the next time step for a given number of species in the current time step.

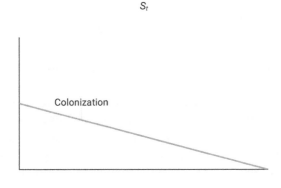

Figure 22.4. Hypothetical change in number of species due to colonization in the next time step for a given number of species in the current time step.

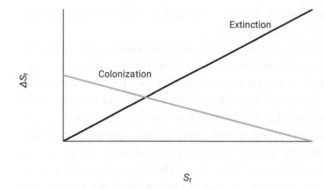

Figure 22.5. Combination of extinction and colonization curves from figures 22.3 and 22.4. The steady-state number of species on the island occurs at the value of S_t where the two curves intersect.

decreases as S_t goes up shows the number of species added to the island due to colonization in the next time step, and the line that increases as S_t goes up shows the number of species removed from the island due to extinction in the next time step.

Looking at the curves together in figure 22.5 makes it clear how a steady-state number of species arises on an island. When there are very few species present, so that S_t is small, more species are added to the island by colonization than are removed via extinction in the next time step. The number of species on the island will thus increase, so that S_{t+1} is greater than S_t. When there are many species on the island, almost the same number as on the mainland, the extinction curve lies above the colonization curve, meaning that more species are removed via extinction in the next time step than are added by colonization. As a result, S_{t+1} will be smaller than S_t. In the middle, where the two curves intersect, the number of species added by colonization is exactly equal to the number removed via extinction. The number of species on the island will thus remain the same into the future. This point represents the steady-state number of species for that island.

In the problem posed in this chapter, there are two islands, one of which is closer to the mainland than the other. Logically, distance from the mainland should be expected to have the most significant impact on the rate at which species colonize the island. Islands that are closer to the mainland would, in general, be expected to receive more colonists in a given time period than islands farther away. The colonization curve for the near island should thus be above that for the far island. However, since both colonization curves must reach zero at the same point, where S_t equals the total number of species on the mainland, the colonization curve for the near island pivots upward from this point, as shown in figure 22.6.

If the only difference between the two islands were distance from the mainland, figure 22.6 shows that the steady-state number of species on the near island should be greater than the number on the far island. This difference is shown graphically by the intersections of the colonization curves with the extinction curve, as the intersection lies farther to the right, at a higher value of S_t, for the near island. However, the problem states that both islands have the same average number of bird species. If the colonization rate is indeed higher on the near island, the only way this can occur is if the extinction rate for the near island is also higher. That situation is shown graphically in figure 22.7.

This logic thus suggests that the number of species going extinct in the next time step at any value of S_t must be larger on the near island than on the far one. As described earlier, within the framework of island biogeography, the extinction rate is thought to increase when species compete more strongly for resources, which makes their populations smaller and more vulnerable to extinction. This can occur when there are more species on an island, which explains the increase in the extinction curve as S_t increases. However, it can also occur when the habitat

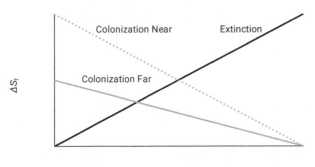

Figure 22.6. Hypothetical increase in colonization curve for the island near the mainland, compared with the island farther from the mainland.

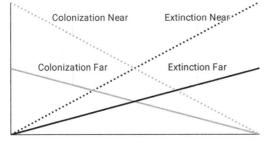

Figure 22.7. Hypothetical increase in extinction curve for the island near the mainland, compared with the island farther from the mainland.

area available for species is smaller, which may generally cause small population sizes to be reached when there are fewer species present. When area is smaller, the extinction curve would be expected to pivot upward from the origin.

For the near island in this example, it thus appears plausible that a higher colonization rate, due to being nearer to the mainland, is balanced by a higher extinction rate, due to available habitat being smaller. This leads both islands to have the same average number of species, as reported in the problem. This conclusion answers the first question posed in the problem, which relates to the available bird habitat on both islands. Our analysis suggests that there is likely to be more bird habitat available on the far island.

The second question asks which of the two islands is likely to experience more turnover in the identities of the bird species present. Figure 22.7 shows that although the average number of species is the same on both islands, the near island experiences more extinction and more colonization at steady state than the far island. If the colonists are a somewhat random selection of species that are found on the mainland but not on the island, this suggests that over any given time step, the identities of the species on the near island are more likely to change than the identities of species on the far island.

It might seem surprising that we can learn so much about the behavior of a system from plots like this, without even performing any calculations. Plots of this nature are often used to explore the behavior of simple models, including several discussed earlier in the book. Difference plots of this same form, for example, are commonly used to understand models based on logistic growth equations, including the Lotka-Volterra competition equations from chapter 3. A plot similar to figure 22.5, for example, can be used to visualize the steady-state population size for the plant species in chapter 2.

NEXT STEPS

Since there were no actual numbers or equations provided in this problem, we had to guess at the particular shapes of the colonization and extinction curves in our figures. In general, the shapes of these curves won't influence our answers to this problem as long as the curves are reasonably smooth, as generally assumed in island biogeographic theory. However, answering more specific questions about the bird communities, such as what fraction of species would go extinct if colonization were cut in half, would require knowing the exact equations describing the curves.

Our solutions to this problem involved the use of discrete-time difference equations. In the original island biogeographic theory, the y-axis of the plot is shown not as a difference, but as a rate, which is based on a continuous-time, differential equation representation of the theory. As is often the case, the broad conclusions from the continuous and discrete time models are the same.

The details of island biogeographic theory can be found in most ecology textbooks, as well as the original works by MacArthur and Wilson (1963, 1967).

23 Max Plant Institute Optimization

A local land trust is preparing to use a recent donation to purchase 1,000 hectares of land for permanent protection. The money could be used to purchase portions of three properties, each representing a different habitat type: wetland, meadow, and forest. The land trust's goal is to divide the purchase across the properties so as to maximize the total number of plant species protected. A biologist at the land trust tells you that the number of species in a protected area commonly increases following the equation $S = cA^z$, where S is the number of species found in some area A, measured in hectares. The parameter c is the number of species found on a single hectare, which the biologist estimates as 50 for the wetland and meadow and 30 for the forest. The parameter z determines how quickly the number of species increases with area. From reviewing the literature, the biologist expects $z = 0.2$ for the wetland and $z = 0.25$ for the meadow and forest. Assuming there is no overlap of species across the properties, how many hectares of land should the trust purchase at each property, and how many species will be protected as a result?

This problem is a familiar one for many conservation organizations throughout the world, which often need to determine how to invest in land purchasing or management to maximize biodiversity protection given a limited budget. The goal of **conservation planning** is to answer such questions in a systematic and quantitative manner.

In many real-world situations, conservation planners have access to maps of species distributions that they can use to select specific areas that will capture as many species as possible. When these maps do not exist, an alternative is to estimate the total number of species that will be found in an area. The equation $S = cA^z$ that describes how species numbers increase with area, known as the **species-area relationship**, has been described as one of the oldest "laws" of ecology. As described in the problem, the parameter c gives the number of species that will be found in a plot of area $A = 1$. This number frequently varies by habitat, as some types of habitat support more species than others in a given area. On the other hand, ecologists have long been interested in the apparent consistency of the parameter z, which is thought to be close to 0.25 for many communities. The reasons behind this law-like behavior are still actively debated.

To begin, we should recognize that the key word in this problem is *maximize*. This word tells us right away that we are facing an optimization problem. In broad terms, optimization is the process of finding the best result from among some set of possible results. We can think of optimization as having three parts, corresponding to three specific words in this definition.

First, the "set" of possible results that we wish to choose from consists of numbers that are calculated using an **objective function**. An objective function is simply an equation that is used to calculate some number of interest. In this problem, we will want to construct an objective function that gives us the estimated total number of plant species protected across the three habitats, which is calculated based on the parameters given in the problem and the area of each habitat that is protected. In other areas of ecology, the objective function is also sometimes called a cost function, fitness function, or loss function, depending on the context of the problem.

Second, the "best" result generally means either the largest or the smallest possible value of the objective function. In this problem, we are interested in maximizing the number of species protected. Our goal is both to determine this maximum number and to determine how much land of the three habitat types must be purchased to attain this maximum.

Third, the set of "possible" results that we can consider is determined by **constraints** that limit the allowable values of the objective function or its parameters. In this problem, for example, we are able to purchase only 1,000 hectares of land. Without this constraint, the way to maximize the number of species protected would be to purchase an infinite amount of land at all three properties!

Let's examine each of these three components of our optimization problem in more detail. To start, we'll focus on the objective function for this problem, which tells us the total number of species protected across the three properties. The total number of species protected, S_T, is simply the sum of the number of species protected at the wetland, meadow, and forest properties, which is

$$S_T = S_W + S_M + S_F$$

We know that the equation $S = cA^z$ will give us the number of species protected at each property. Using the particular values of c and z for each property, we can write this equation as

$$S_T = 50(A_W)^{0.20} + 50(A_M)^{0.25} + 30(A_F)^{0.25}$$

There are three unknown parameters in the equation above that are needed to calculate S_T, which are the numbers of hectares to be purchased in each of the three habitats. Our goal is to modify these three numbers in a way that maximizes the value of S_T. When doing so, however, we have to follow the constraint that

$$A_W + A_M + A_F = 1{,}000$$

We also have the implicit constraint that none of the three areas can be negative.

Now that we have determined our objective function and constraints, we can move on to solving our optimization problem. Perhaps the simplest way to solve an optimization problem is through **direct search**. This method involves simply calculating the value of the objective function for all possible values of the parameters. In this case, for example, we might consider integer numbers of hectares that could be purchased at each property, from 0 to 1,000, trying all possible combinations of every number for each property. This simple approach, however, would give us 1 billion possible values of S_T to calculate, most of which would not respect the constraint that the total number of hectares purchased across all three properties must be equal to 1,000.

Aside from direct search, there are many other ways to solve optimization problems, including the use of calculus to calculate derivatives and gradients, iterative methods that sequentially step closer to nearby minima or maxima, and stochastic methods that perform a search for minima and maxima with some element of randomness. Here we'll use a simple variation of a **random search** algorithm, which falls into this third category.

To use this random search algorithm, we'll start by choosing initial values for our parameters A_W, A_M, and A_F. We'll start with the values of $A_W = 800$, $A_M = 100$, and $A_F = 100$, which reflect an initial strategy of preferentially purchasing land at the property with the highest number of species on a single hectare. Notice that these three parameter values respect the constraint that the total area purchased must be 1,000 hectares. Plugging these three values into our objective function shows that with this combination of habitats, we could protect a total of 443 species.

To use the random search algorithm, we begin with a set of current values for the three parameters, which we'll call our current point. When we start our algorithm, our current point consists of $A_W = 800$, $A_M = 100$, and $A_F = 100$, which leads to the protection of 443 species. We then choose a range of possible values for the parameters that are all relatively close to their current values—within 10 hectares of their current values, for example. Within this allowable range, we then choose a random value for each of the parameters. For example, in our implementation of this algorithm below, we'll choose a random number between −10 and 10 and add this number to the current number of hectares protected at each property. We might, for example, randomly select the values $A_W = 794$, $A_M = 107$, and $A_F = 99$. This new set of values forms our candidate point.

Next, we calculate the value of the objective function at this candidate point and compare it with the value of the objective function at the current point. In this case, our candidate point would protect a total of 446 species. If the value of the objective function at the candidate point is greater than that at our current point, we move to it, and if it's lower, we stay with our current point. In this case, 446 is greater than 443, so we replace our current point values with the values from the candidate point, then continue to the next step. We repeat this procedure for many steps, updating our current point as needed.

In essence, the idea of the random search algorithm is to explore a small neighborhood of possible values around our current point in each step, moving to better points as they are found. Eventually, we'll move to better and better points until we become stuck at the same current point for a long enough time that we believe we have found the maximum value of the objective function, or a value very close to it.

As we actually implement this random search algorithm for this problem, we'll need to add one more slight complication. If we were to choose a random nearby value for each of the three habitats to form a candidate point, it's very likely that the total number of hectares protected across all three properties would no longer equal exactly 1,000 hectares. In order to respect the constraint that the total area protected across all three properties is 1,000 hectares, we will instead choose random values for the area of wetland and meadow in our candidate point and then calculate the area of forest that is needed so that the total area protected in our candidate point is still 1,000 hectares.

> In the first row of a new Google Sheet, in the first nine columns, enter the text Aw, Am, Af, St, Aw-candidate, Am-candidate, Af-candidate, St-candidate, and Move?. The first four columns represent our current point. In Cells A2, B2, and C2, enter the numbers 800, 100, and 100, corresponding to our initial guesses for the area to purchase at each property. In Cell D2, enter the formula

=50*A2^0.2+50*B2^0.25+30*C2^0.25, which uses our objective function to calculate S_T for our current point.

The next five columns represent a new candidate point. In Cells E2 and F2, enter the formulas =A2+randbetween(-10,10) and =B2+randbetween(-10,10), which will add a random integer between -10 and 10 to the current values of A_W and A_M. In Cell G2, enter the formula =1000-F2-E2, which will give the candidate value of A_F while ensuring that the sum of the area for the three properties equals 1,000. In Cell H2, enter the formula =50*E2^0.2+50*F2^0.25+30*G2^0.25 to calculate S_T for the candidate point, just as we did for the current point. Finally, in Cell I2, enter the formula =H2>D2, which will return True if the candidate point protects more species than the current point.

Next we make a decision in Row 3 about whether to move to the candidate point. We do this if the candidate point protects more species than the current point. In Cell A3, enter the formula =if($I2,E2,A2), which updates the current value of A_W with the candidate value only if the candidate S_T is greater than the S_T in the previous row. Highlight this cell and drag it to the right to Column D to make the same decision for the other two areas and S_T.

Now we're ready to perform the repeated random search. Highlight Cells A3 to D3 and drag them down to Row 500. Then highlight Cells E2 to I2 and drag those down to Row 500. You will see the current point begin to move quickly toward the true maximum, and then slowly settle on values very close to or equal to the answer given below.

Figure 23.1 shows the area of each habitat protected during each step in our random search algorithm. Notice that the areas start with our initial values and change at a fairly steady rate before eventually settling into a nearly steady state after about 300 steps. After 500 steps, the areas protected are $A_W = 254$, $A_M = 495$, and $A_F = 251$. With these values, the total number of species protected is $S_T = 507$, which is the maximum number that can be protected with any combination of habitat from the three properties.

NEXT STEPS

In this problem, the choice of starting values for our random search algorithm does not matter very much, as the random search algorithm will always be able to find the true maximum for this objective function. In other optimization problems, however, it can be important to choose

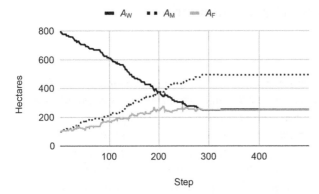

Figure 23.1. Area of wetland (A_W), meadow (A_M), and forest (A_F) protected at each step in the optimization algorithm. After about 300 steps, the algorithm finds a nearly optimal combination of habitats to maximize total species richness.

initial values that are as close to the maxima or minima as possible. Formally speaking, this is because the objective function in this problem is **convex**, which means that it is shaped like an upside-down bowl with a single maximum that you can get to by "climbing" the sides of the bowl from any starting point. The most difficult optimization problems are ones with **local minima** or maxima. In these cases, if you simply attempt to increase or decrease the objective function step by step, you may get stuck at a low or high point that is not the absolute lowest or highest point that can be found with any combination of parameters.

As described above, the species-area relationship is one of the oldest and most widely studied ecological patterns, and you can find more information on this relationship in many textbooks. Rosenzweig (1995) provides a comprehensive overview of this relationship and its possible interpretations.

24 **Bears with Me** Stochastic Simulation

> *You read a newspaper article about a small remnant population of bears in a national park. This population has been monitored for the past 11 years, and the adult female bear population in each successive year has been counted at 40, 39, 39, 42, 39, 41, 40, 33, 36, 34, and 39. Biologists believe that if the population were ever to fall to a threshold of 10 adult females, it would be functionally extinct and extremely unlikely to recover. Given this pattern of year-to-year variation in the population, what is the probability that this bear population will go functionally extinct in the next 100 years?*

This problem asks us to estimate the risk of extinction for a population, given some knowledge about how population size changes from year to year. A class of models known as **population viability analyses** are often used to estimate extinction risk when the year-to-year change in population size is believed to have some element of randomness to it. When the only information available about a population is a count, as in this problem, the appropriate model is known as a count-based population viability analysis. Since we are projecting a population into the future while accounting for some amount of randomness, this analysis will incorporate ideas from part 1 and part 2 of this book in a single model.

Unlike previous chapters in this book that dealt with extinction, this problem suggests that there is a population size greater than zero at which the population becomes functionally extinct. This size is known as a **quasi-extinction threshold**. Functional extinction occurs when a population still exists but is thought to be in a "zombie" state that is inevitably headed toward eventual extinction. This can occur when the population becomes so small that individuals have difficulty finding mates or mounting group defenses against predators, or become susceptible to random events that can wipe out the entire population.

Let's start our work on this problem by looking at the change in population size from year to year. Population counts are shown in the top row of table 24.1. By dividing the size of the population in each year by its size in the previous year, we can calculate a set of 10 numbers. Each of these numbers represents the factor by which the population grew or shrank between one year and the next. In chapter 2, this factor was named R, while later, in part 3 of the book, it was called λ. For consistency with

Table 24.1 Change in bear population from year to year (λ)

N_t	40	39	39	42	39	41	40	33	36	34	39
λ		0.975	1.000	1.077	0.929	1.051	0.976	0.825	1.091	0.944	1.147

other population viability analyses, which are often based on matrix models, we'll call this factor λ.

The bottom row of table 24.1 gives the 10 values of λ that were observed for this population across the 11 years of monitoring. We can see from these calculations that the population grows in some years, corresponding to $\lambda > 1$, and decreases in other years, corresponding to $\lambda < 1$. If we calculate the average value of λ by adding up these 10 numbers and dividing by 10, we find that this average is equal to 1.001. We might take this to mean that in an "average" year, the population increases very slightly, and that, as a result, the population should be safe from extinction in the long term.

To actually calculate the risk of extinction for this population, however, we'll need a model to project population sizes into the future. In the absence of any other information about the population, we might begin by turning to the simple exponential population growth model from chapter 2. This model projects the population change from year to year as

$$N_{t+1} = \lambda_t N_t$$

In all of our population projections in parts 1 and 3 of this book, we had only a single, constant yearly growth rate that we used for every time period, which was represented by the parameter λ. The data on this bear population, however, make it clear that its λ is not constant, but instead varies from year to year. In the model above, we thus use the parameter λ_t to indicate the yearly growth rate for the population in a particular year t. This variation is potentially very important. It's conceivable, for example, that a string of several bad years in a row might be enough to drive the population below the quasi-extinction threshold, even if, over the long term, the population has an average growth rate that is positive.

The next step is to determine how we will choose λ_t for each of the next 100 years, the time period over which we want to project our population. To do this, we'll consider the value of λ_t in each year to be a random variable that we draw from some probability distribution. Each year, for 100 years, we'll randomly draw a value of λ_t and use that number to calculate N_{t+1} from N_t. We will expect to see the population increase in some years and decrease in others. Our goal will be to determine whether at any point in the next 100 years, the population reaches the quasi-extinction threshold of 10 adult females.

The procedure above will generate a single **stochastic simulation** of the bear population for the next 100 years. If we repeat this simulation many times, we can then examine the fraction of the simulations in which the population reaches the quasi-extinction threshold, which we can interpret as the probability that any single randomly chosen simulation will lead to extinction. This will be our estimate of the probability that our one real population reaches this extinction threshold in the next 100 years.

To perform these calculations, we must first choose the probability distribution from which we will draw our random values of λ_t. One possible approach here is to apply the methods that we discussed in part 4. This would involve choosing a specific probability distribution to describe λ_t, using maximum likelihood methods to fit this distribution to the 10 values of λ_t that we've observed, and then drawing a random number from this distribution to represent λ_t in each year. Here, however, we'll take the simpler approach of using a probability distribution that consists simply of the 10 historical values of λ, each assigned a probability of 0.1 of occurring in a future year. In other words, for each year of each simulation, we'll randomly choose one of the values of λ_t from table 24.1 and multiply this chosen value by N_t in order to calculate N_{t+1}. The advantage of this approach is that we do not have to make any assumptions about the true probability distribution from which these counts are drawn, as we did in part 4. We'll perform 100 simulations of the bear population, each for 100 years, using this approach.

In Column A of a new Google Sheet, enter the word lambda in Cell A1, followed below by the 10 values of λ from table 24.1. Then, in Column C, enter the variable name t, followed by the numbers 0 to 100 in successive rows. These are the years of each simulation. In the next 100 columns, from Column D through Column CY, enter the number 39 in Row 2. This represents the starting population for each simulation. To add columns, highlight the existing columns in the sheet and right-click on any column name, then choose Insert Columns.

Then, in Cell D3, the first year of the first simulation, enter the formula =D2*offset($A$2,randbetween(0,9),0). This formula will randomly select one of the 10 values of λ from Column A and multiply the population in the previous year by this number. Drag the bottom right corner of this cell down to Row 101 to simulate the first complete population trajectory. Drag the bottom right corner of Cell D101 to the right to Cell CY101 to create 100 simulated populations.

Figure 24.1 shows the results of 10 simulated trajectories for this bear population. Because of the random nature of each simulation, your simulated trajectories will look similar, but not identical, to those in this figure. We can see that in some cases, the population experiences a series of good years and increases. However, in some simulations, the population experiences a series of bad years and shrinks. In two of the simulations shown in figure 24.1, the population reaches 10 individuals, our quasi-extinction threshold, at some point during the 100 years. By the criteria in this problem, those populations should be marked as functionally extinct at that point and the simulation ended.

To indicate when a simulated population becomes functionally extinct, modify the formula in Cell D3 to read =if(D2<=10,0, D2*offset($A$2,randbetween(0,9),0)). Drag the bottom right corner of this cell down and to the right to Cell CY101 to replace all formulas with this new formula. This new formula will record the population in N_t as 0 if the previous year's population reached the quasi-extinction threshold. This population size of 0 will then repeat from that year until year 100. The proportion of 0s in year 100 represents the probability that a single simulated population will reach the quasi-extinction threshold at some point during the next 100 years.

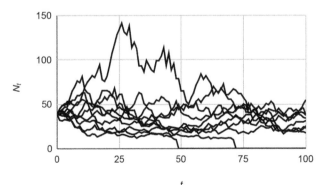

Figure 24.1. Ten simulated trajectories for the bear population over the next 100 years. In two of the ten simulations, the population reaches the quasi-extinction threshold before year 100.

In 16 out of my 100 simulations, the population went functionally extinct at some time before year 100. Your estimate may be different due to the randomness of the simulations, but should be close to this number. We can thus estimate that, based on our model, there is a 16% chance that the one real bear population of this national park will go extinct in the next 100 years.

NEXT STEPS

Stochastic simulation techniques, such as the one in this chapter, are very widely used for projecting populations into the future, particularly over long time periods. The utility of this approach comes from its ability to combine a deterministic population growth model, like the ones from part 1 of this book, with randomness or uncertainty in the parameters of that growth model. These types of models are encountered particularly frequently in conservation assessments of species at risk.

The data used in this problem are taken from 11 years of population censuses of the grizzly bear population of Yellowstone National Park in the United States (reprinted in Morris and Doak 2002). This population was the subject of one of the first population viability analyses that was ever published (Shaffer 1983). See these two references for a discussion of other, more complex models that can be applied to these data and to other similar data sets. While Morris and Doak use a quasi-extinction threshold of 20 individuals, we've here used the more conservative threshold of 10 individuals.

Incidentally, you might be interested to know that the real grizzly bear population of Yellowstone National Park has not gone extinct, although several historical population viability analyses predicted a relatively high probability of extinction by today.

25 **Natives in the Neighborhood** Cellular Automata

You join a volunteer group that is restoring a native plant species along a small section of riverbank in a local park. The species is a large plant, and only about 50 plants at most could fit side by side along the bank. The goal is to plant 10 evenly spaced individuals this year, with the hope that this initial population will expand to cover the entire riverbank. You decide to build a model to see how long this will take. You begin by imagining the riverbank as a set of 50 cells, side by side, 10 of which will have a plant initially. Since the plant produces large fruits that are unlikely to disperse far, an empty cell with no neighbors is very unlikely to be colonized in the next year, but an empty cell with a plant in a neighboring cell is very likely to be colonized. A cell that is already occupied will remain occupied unless both of its neighboring cells are also occupied, in which case the "middle" plant is likely to die due to crowding. How many years will it take until the riverbank is full of plants?

In its initial setup, this problem appears similar to the population growth models that we discussed in part 1. As in chapter 3, we have an initial population and a carrying capacity, in this case 10 and 50 individuals, respectively. Our goal is to project the change in the total population size into the future.

Although our goals in this chapter and in chapter 3 are essentially the same, the model that is described in this problem is quite different from those in part 1. Instead of setting up a difference equation to track the total number of individuals from year to year, we will instead choose a model that tracks the presence or absence of a plant in a specific cell in a given year. We need to do this in order to incorporate the spatial information about colonization and death that is given in the problem. Specifically, we need to be able to enforce the restrictions that a cell can be colonized only if one of its neighboring cells is occupied and that a cell will become empty if both of its neighboring cells are occupied. There is no way to incorporate this spatial information into the models from part 1, which track only the total population size from year to year.

The model described above is an example of a **cellular automaton**. A cellular automaton is a cell-based model in which the state of each cell in the next time period $t + 1$ depends on the state of the cell and its nearest neighbors at time t. The simplest of these models have only two states for a cell, which we will call occupied and unoccupied. These models are also particularly simple when they have cells that all lie in a row, making the model one-dimensional, such that we can determine the fate of any cell by looking at only two other cells, the one to its left and the one to its right. We'll use a cellular automaton to project the native plant population on the riverbank into the future.

The problem provides almost all of the information that we will need to project whether each cell on the riverbank is occupied or unoccupied from one year to the next. However, we need to decide what to do about the cells at the end of the riverbank, which have a neighbor on only one side. We'll assume that if the one neighbor cell is occupied, these end cells will become colonized, just like the cells in the middle. Since these end cells can never have two neighbors, we'll assume that once they are colonized, they will never become empty again due to crowding. With those extra rules in hand, we can proceed to calculate how the population of native plants will change over time.

In a new Google Sheet, highlight the first 50 columns (Columns A through AX) by clicking on the letter A in the first column header, holding Shift, then clicking on the letters AX at the top of Column AX. Drag the right edge of Column AX to the left to make the column thin, so that all 50 columns will be visible on your screen at once. Each column will represent one of our 50 cells, and each row will represent a year of the native plant population projection. Cells with a value of 1 will be occupied, and cells with a value of 0 will be unoccupied.

In Row 1, enter the numbers 1, 0, 0, 0, 0 in the first five columns, then repeat these five values 10 more times. These values represent the set of 10 evenly spaced individuals that are planted in the first year. In the second row, in Cell A2, enter the formula =if(sum(A1:B1)>0,1,0). This formula applies to an edge cell, making the cell a 1 if either the cell itself or its neighbor is a 1 in the previous time step. Similarly, enter the formula =if(sum(AW1:AX1)>0,1,0) in Cell AX2.

For the middle cells, enter the formula =if(OR(sum(A1:C1)=1,sum(A1:C1)=2),1,0) in Cell B2. It may take a little bit of thinking to see how it does so, but this formula enforces all of the colonization and death rules described in the problem. It does this by making a cell occupied if either one or two of the cell and its neighbors were occupied in the previous time step. Only if all three cells were either occupied or unoccupied does the cell become unoccupied in the next time step. Drag the corner of Cell B2 to the right to Cell AW2. Then highlight all 50 cells in Row 2 and drag these down for 200 rows, representing 200 time steps.

To see the pattern of occupied and unoccupied cells more clearly, highlight the first 50 columns, then go to Format → Conditional formatting in the Menu bar. Here, change the box under *Format cells if...* to *Is equal to*, then in the box labeled *Value or format*, type the number 1. Choose a color for the cells containing a 1 to make the pattern easy to see.

When we actually project this native plant population into the future, you'll see something that you probably did not expect. As you move down the sheet, the number of occupied cells does not increase smoothly until the entire landscape is colonized, nor does the entire landscape become empty, nor does the population reach an obvious steady state in between. Cells seem to become empty and colonized from year to year following a very strange, somewhat repetitive pattern that is hard to describe in words. Figure 25.1 shows occupied cells in dark colors and unoccupied cells in light colors for the first 50 years of our population projection. The 50 cells in the landscape run from left to right in rows, and time runs into the future going down the sheet.

What appears to happen is that occasionally, a whole section of riverbank becomes occupied in some year, only to become completely empty the following year as all of the plants in this section die due to crowding. This large empty space then slowly fills up again by colonization from the edges until the area is once again full. Just as soon as that happens, however, another large empty space opens up somewhere else.

Despite the very unusual spatial pattern shown in figure 25.1, we might wonder if there could be some underlying regularity in the pattern. For example, perhaps the total number of cells that are occupied at any given time step reaches some approximate steady-state number, even if their positions change, or perhaps the number of occupied cells cycles between a few different numbers.

Figure 25.2 shows the fraction of cells that are occupied at each time step in our calculation. It appears that the number of occupied cells never settles down to a stable number between 0 and 50. Instead, the number of occupied cells changes seemingly at random from year to year, with roughly 40%–60% of the riverbank occupied in most years.

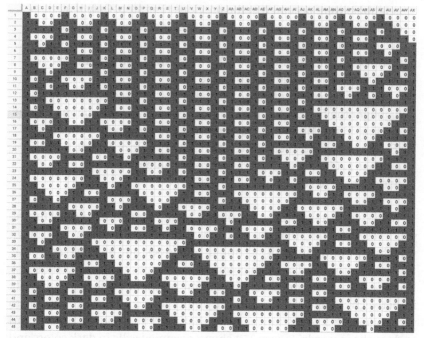

Figure 25.1. The set of occupied cells on the riverbank (horizontal dimension) over time (vertical dimension).

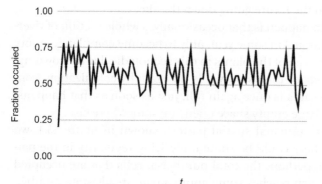

Figure 25.2. The fraction of cells that are occupied over time.

Figure 25.2 provides a clear, if unusual, answer to the question posed in the problem. It turns out that the landscape will never be entirely full of native plants, no matter how long we wait. Enough cells become empty each year, through death by crowding, that the colonization process can never entirely fill up the landscape. However, the number of plants also doesn't reach some steady-state number below its upper bound of 50. Rather, the population is highly unstable from year to year.

This model seems to behave very differently from all of the ones that we've seen so far in this book, and we might wonder what exactly is going on here and what to make of it. There are several differences between this model and the ones that we've seen before, but perhaps the most important difference is that this model is **computational** in nature, rather than mathematical. In other words, we cannot write down a simple equation that shows how the whole population changes over time. Instead, we have written down a very simple set of rules that tracks the births and deaths of individual organisms by the state of their neighbors. This seemingly very simple set of rules has led to strangely complex behavior at the level of the whole landscape.

This emergence of complexity from simplicity is a general theme that appears in many rule-based computational models, particularly those in which space and adjacency are involved. It turns out that it is often impossible to solve for the long-term or large-scale behavior of the model using mathematical tools, and the model must instead be "solved" by simply applying the rules to see what happens using a simulation. These types of simulated, rule-based models are continuing to grow in importance in quantitative ecology, and there is still much room for exploring and applying models like these to ecological questions.

NEXT STEPS

As mentioned above, this computational model differs in several ways from the other models that we've encountered so far in this book. One difference that may be less obvious is that this model turns out to be very sensitive to initial conditions. Try, for example, modifying the values in Row 1 of the sheet to read 0, 0, 1, 0, 0 repeated 10 times. You'll now see that a regular pattern emerges in which, in the long term, the number of plants fluctuates back and forth between two values over and over again. A deterministic model like this one in which all projections are made exactly, with no randomness, but in which the projections are highly sensitive to initial conditions, is said to exhibit the hallmarks of **chaos**.

Cellular automata are one type of model within a much larger class of computational models. Aside from their use in ecology, such models are often discussed under the heading of complexity theory or complex systems science. Mitchell (2009) provides a generally readable introduction to this overall field. Wolfram (2002) is a large, somewhat controversial, somewhat confusing, and mostly definitive guide to cellular automata.

The specific cellular automaton rule set that we used here was classified by Wolfram as Rule 126. If you find these models interesting, you might also be interested in looking into Joseph Conway's Game of Life, which is a two-dimensional cellular automaton with a set of rules that leads to very rich and interesting behavior.

References

Allen, Linda J. S. 2010. *An Introduction to Stochastic Processes with Applications to Biology*. Boca Raton, FL: CRC Press.

Anderson, Roy M., Helen C. Jackson, Robert M. May, and Anthony M. Smith. 1981. "Population Dynamics of Fox Rabies in Europe." *Nature* 289: 765–71.

Bolker, Benjamin M. 2008. *Ecological Models and Data in R*. Princeton, NJ: Princeton University Press.

Burnham, Kenneth P., and David R. Anderson. 2013. *Model Selection and Inference: A Practical Information-Theoretic Approach*. New York: Springer Science & Business Media.

Caswell, Hal. 2006. *Matrix Population Models*. Sunderland, MA: Sinauer Associates.

Day, Jemery R., and Hugh P. Possingham. 1995. "A Stochastic Metapopulation Model with Variability in Patch Size and Position." *Theoretical Population Biology* 48: 333–60.

Diggle, Peter J. 2013. *Statistical Analysis of Spatial and Spatio-Temporal Point Patterns*. Boca Raton, FL: CRC Press.

Green, Jessica L., and Joshua B. Plotkin. 2007. "A Statistical Theory for Sampling Species Abundances." *Ecology Letters* 10 (11): 1037–45. https://doi.org/10.1111/j.1461-0248.2007.01101.x.

Harte, John. 2011. *Maximum Entropy and Ecology: A Theory of Abundance, Distribution, and Energetics*. Oxford Series in Ecology and Evolution. Oxford: Oxford University Press.

Horn, Henry S. 1975. "Markovian Properties of Forest Succession." In *Ecology and Evolution of Communities*, edited by M. Cody and J. Diamond, 196–211. Cambridge, MA: Belknap Press.

Kling, Rakel, Eleni Galanis, Muhammad Morshed, and David M Patrick. 2015. "Diagnostic Testing for Lyme Disease: Beware of False Positives." *BC Medical Journal* 57 (9): 396–99.

Kremer, William. 2014. "Do Doctors Understand Test Results?" *BBC News*, July 7, 2014, sec. Magazine. https://www.bbc.com/news/magazine-28166019.

Leslie, P. H. 1945. "On the Use of Matrices in Certain Population Mathematics." *Biometrika* 333 (3): 183–212.

MacArthur, Robert H., and Edward O. Wilson. 1963. "An Equilibrium Theory of Insular Zoogeography." *Evolution* 17 (4): 373–87.

MacArthur, Robert H., and Edward O. Wilson. 1967. *The Theory of Island Biogeography*. Princeton, NJ: Princeton University Press.

May, Robert. 1975. "Patterns of Species Abundance and Diversity." In *Ecology* and *Evolution of Communities*, edited by M. Cody and J. Diamond, 196–211. Cambridge, MA: Belknap Press.

McGahan, Jerry. 1968. "Ecology of the Golden Eagle." *Auk* 85 (1): 1–12. https://doi.org/10.2307/4083617.

Mitchell, Melanie. 2009. *Complexity: A Guided Tour*. Oxford: Oxford University Press.

Morris, William F., and Daniel F. Doak. 2002. *Quantitative Conservation Biology: Theory and Practice of Population Viability Analysis*. Sunderland, MA: Sinauer Associates.

Otto, Sarah P., and Troy Day. 2007. *A Biologist's Guide to Mathematical Modeling in Ecology and Evolution*. Princeton, NJ: Princeton University Press.

Rosenzweig, Michael L. 1995. *Species Diversity in Space and Time*. Cambridge: Cambridge University Press.

Shaffer, Mark L. 1983. "Determining Minimum Viable Population Sizes for the Grizzly Bear." *Bears: Their Biology and Management* 5: 133–39. https://doi.org/10.2307/3872530.

Strauss, David. 1975. "A Model for Clustering." *Biometrika* 62 (2): 467–75.

Whitlock, Michael C., and Dolph Schluter. 2019. *The Analysis of Biological Data*. New York: Macmillan Higher Education.

Wolfram, Stephen. 2002. *A New Kind of Science*. Champaign, IL: Wolfram Media.

Zuur, Alain F., ed. 2009. *Mixed Effects Models and Extensions in Ecology with R*. Statistics for Biology and Health. New York: Springer.

Index

The letter *f* following a page number denotes a figure.

absorbing state, 82–86, 96
absorption: colonization and, 81–84, 87; extinction and, 81–87; Google Sheets and, 85; matrices and, 81–82, 85–87; metapopulations and, 81–86; probability and, 81–86; time steps and, 86; transition matrices and, 82, 85; vectors and, 85
age-structured models: beetles and, 69–75; difference equations and, 70; Google Sheets and, 72; matrices and, 74–75; projections and, 70–71, 75; time steps and, 71–72; vectors and, 71–75
Akaike's information criterion (AIC), 113–14
algebra, 2, 19, 26, 66, 84
algorithms, 55, 142–45
Allen, Linda J. S., 97
Anderson, Roy M., 32, 113–14

base rate fallacy, 55
Bayesian probability: base rate fallacy and, 55; contingency tables and, 50–52; random variables and, 37; ticks and, 50–56
Bayes's rule, 3, 53–56
bears, 146–50
beetles, 69–75
behavioral ecology, 1
bias, 101–2
binomial distribution, 110, 114
birds: egg counts, 38–49, 103; graphical thinking and, 132–39; island biogeography and, 132–39; single-variable probability and, 38–42; two-variable probability and, 43–49
birth-death chains, 96–97
birth rates, 16–17
blocking, 42
boundary conditions, 96

Burnham, Kenneth P., 113–14
butterflies, 81–87

calculus, 2, 10, 32, 142
carrying capacity, 19–20, 26, 151
Caswell, Hal, 75
cellular automata: carrying capacity and, 26; cell states and, 151–55; chaos and, 155; colonization and, 151–54; Conway and, 156; deterministic component of, 155; difference equations and, 151; Game of Life and, 156; initial conditions and, 155; plants and, 151–56; projections and, 152–53, 155; randomness and, 151; Rule 126 and, 156; time steps and, 152–53; Wolfram and, 155–56
chaos, 155
closed populations, 16
colonization: absorption and, 81–84, 87; cellular automata and, 151–54; extinction and, 132 (*see also* extinction); island biogeography and, 132–39; logistic growth and, 16; Lotka-Volterra competition and, 22–26
column stochastic matrices, 79–80, 87
community, 1, 22, 27, 76, 78
competition coefficients, 26
competitive exclusion, 24, 26
computational models, 1, 118, 131, 155
conditional probability, 47–55
conservation planning, 15, 131, 140, 150
Consider a Spherical Cow (Harte), 4
constraints, 141–42
contingency tables: Bayesian probability and, 50–52; two-variable probability and, 43–49
convex functions, 145
Conway, Joseph, 156
cycles, 76, 153

159

Day, Jemery R., 86–87
Day, Troy, 10, 27, 97
deterministic models, 117–18, 120, 123, 150, 155
difference equations: age-structured models and, 70; basics of, 2–3, 7–10; cellular automata and, 151; duckweed and, 11–12, 15; exponential growth and, 11–12, 15; Google Sheets and, 9; graphical thinking and, 133, 139; initial conditions and, 8; logistic growth and, 18–19; Lotka-Volterra competition and, 23–26; matrices and, 65, 76–78; seeds and, 7–10; SIR models and, 29–32; time steps and, 7–10; variables and, 7
diffusion: birth-death chains and, 96–97; boundary conditions and, 96; distribution and, 95–96; with drift, 96; extinction and, 96; fish and, 88–97; Google Sheets and, 94; metapopulations and, 88, 90; probability and, 89–96; transition matrices and, 88–97; vectors and, 88–95
Diggle, Peter J., 109
direct search, 142
distribution: basics of, 35–36; binomial, 110, 114; diffusion and, 95–96; discrete, 36, 61, 96; eigenvectors and, 75; generalized linear models, 115–20; geometric, 58–61, 110–11, 114; Google Sheets and, 59; hypothesis testing, 121–28; maximum likelihood, 103–9; model selection and, 110–14; null, 3, 123–28; optimization, 140; Poisson, 105–24; projections and, 3, 57, 75; rabbit populations and, 57–61; randomness and, 35 (*see also* randomness); single-variable probability, 39–42; stable age, 74–75, 78; stationary state and, 79; stochastic simulation, 147–48; support of, 35, 39, 60–61, 104–5, 140; time steps and, 57, 61, 90–96; transition matrices and, 78–79
dominant eigenvalue, 74–75

duckweed: difference equations and, 11–12, 15; exponential growth and, 11–15; logistic growth and, 16–17; matrices and, 65–68; probability and, 35

ecosystems, 1, 14
egg counts, 38–49, 103
eigenvalues, 74–75
eigenvectors, 75
epidemics, 28–30
epidemic threshold, 31
error, 35, 50, 128
expected value, 41
exponential growth: difference equations and, 11–12, 15; duckweed and, 11–15; Google Sheets and, 12–13; prediction and, 11; projections and, 11, 13–14; reproductive factor, 14–15, 74; time steps and, 12, 14–15
extinction: absorption and, 81–87; diffusion and, 96; graphical thinking and, 132–38; island biogeography and, 132–38; population viability analyses and, 146–47, 150; quasi-extinction threshold and, 146–50; stochastic simulation and, 146–50

fish: diffusion and, 88–97; release point and, 88–89, 94–96
fitting the model, 106
forests: generalized linear models and, 115; optimization and, 140–43, 145f; population samples and, 101–2; seedling counts and, 103–14; statistics and, 101–2; succession and, 76–80; transition matrices and, 76–80
foxes, 28–32
frequentist approach, 36–38, 56, 79, 101–3, 121
frogs: generalized linear models and, 115–20; hypothesis testing and, 121–28; Poisson distribution and, 115–24; roadkill counts, 115–28

Game of Life, 156
generalized linear models (GLMs), 102; forests and, 115; frogs and, 115–20; log likelihood and, 118–19;

maximum likelihood and, 118–19; optimization and, 118–19; Poisson distribution and, 115–20; probability and, 115–16, 120; randomness and, 118; seeds and, 115; stochastic simulation and, 117–18, 120
geometric distribution, 58–61, 110–11, 114
geometric growth, 15, 59
Google Sheets: absorption and, 85; age-structured models and, 72; difference equations and, 9; diffusion and, 94; distribution and, 59; exponential growth and, 12–13; hypothesis testing and, 124, 126; instructions for, 3; logistic growth and, 18, 21; Lotka-Volterra competition and, 24; maximum likelihood, 107; model selection and, 111; optimization, 143; SIR models, 30; stochastic simulation, 148, 152; transition matrices, 78
graphical thinking: birds and, 132–39; difference equations and, 133, 139; extinction and, 132–38; time steps and, 133–38
Green, Jessica L., 114

Harte, John, 4, 114
Horn, Henry S., 80
hypothesis testing: deterministic component of, 117–18, 120, 123; frogs and, 121–28; Google Sheets and, 124, 126; likelihood ratio test and, 128; maximum likelihood and, 121, 124–25; null distribution and, 3, 123–28; parametric bootstrapping and, 122–24, 128; Poisson distribution and, 121–24; probability and, 126–28; P-value and, 123, 126–27; randomness and, 125; stochastic simulation and, 123; test statistic and, 122–28; t-test and, 120, 127–28; Wald test and, 128

independence, 36, 39–42, 48–49, 58
infection, 20, 28–32, 50
initial conditions, 8, 155
interactions, 43

interspecific competition, 22–23, 27
intraspecific competition, 17, 20
island biogeography, 132–39

joint probability, 46–49, 53, 58

Kling, Rakel, 55
Kremer, William, 50

landscape ecology, 1, 28, 38, 45, 81, 153–55
law of total probability, 54
Leslie, P. H., 75
Levins metapopulation model, 20
likelihood ratio test, 128
linear algebra, 66
linear equations, 66, 72
local minima, 145
logarithms, 106, 120
logistic growth: birth rates, 16–17; carrying capacity and, 19–20, 26, 151; closed populations, 16; colonization and, 16; difference equations, 18–19; duckweed and, 16–17; Google Sheets and, 18, 21; intraspecific competition, 17, 20; Lotka-Volterra competition, 26–27, 138; metapopulations and, 20; open populations and, 16; phenomenological approach and, 20; post-breeding census, 21; pre-breeding census, 21; projections and, 16, 21; randomness and, 18, 20; seeds and, 16–21; shade and, 16–21; SIR models and, 32; time steps and, 21
log likelihood, 106–9, 111–13, 118–19
Lotka-Volterra competition: carrying capacity and, 26; colonization and, 22–26; competition coefficients and, 26; competitive exclusion, 24, 26; difference equations and, 23–26; food webs and, 27; Google Sheets and, 24; interspecific competition, 22–23, 27; logistic growth and, 26–27, 138; phenomenological approach and, 27; prediction and, 23; projections and, 24; seeds and, 22–23, 26; shade and, 22–27

MacArthur, Robert H., 139
macroecology, 1, 114
marginal probability, 46, 48, 53–54
matrices, 2–3; absorption and, 81–82, 85–87; age-structured models and, 70–75; basics of, 64–68; column stochastic, 79–80, 87; difference equations and, 65, 76–78; diffusion, 88–96; duckweed and, 65–68; eigenvectors and, 75; exponential growth and, 15; linear equations and, 66, 72; Poisson distribution and, 110–14; projection, 70–71, 75, 77; row stochastic, 79–80, 87; samples and, 110–14; stochastic simulation and, 147; transition, 76–82, 85, 88–96; vectors and, 66–68, 71–79, 85, 88–95
matrix multiplication, 67, 72, 78
maximum likelihood: Akaike's information criterion (AIC) and, 113–14; estimate of, 106–8, 111–13, 118–19, 121, 124–25; generalized linear models and, 118–19; Google Sheets and, 107; hypothesis testing and, 121, 124–25; log likelihood and, 106–9, 111–13, 118–19; model selection and, 111, 113; optimization and, 109; Poisson distribution and, 105–9; probability and, 103–9; randomness and, 104; seeds and, 103–9; stochastic simulation and, 148
May, Robert M., 114
McGahan, Jerry, 42
mean value, 41
metapopulations: absorption and, 81–86; diffusion and, 88, 90; Levins model of, 20; logistic growth and, 20
Mitchell, Melanie, 155
model selection: Akaike's information criterion (AIC) and, 113–14; Google Sheets and, 111; log likelihood and, 111–13; maximum likelihood and, 111, 113; Poisson distribution and, 110–14; prediction and, 110–11, 114; probability and, 110, 112; seeds and, 110–14

null distribution, 3, 123–28
null hypothesis, 122–28

objective function, 141–45
open populations, 16
optimization: conservation planning and, 140; constraints and, 141–42; convex functions, 145; direct search and, 142; forests and, 140–43, 145f; generalized linear models, 118–19; Google Sheets and, 143; local minima and, 145; maximum likelihood, 109; objective function and, 141–45; principles of, 1, 3, 131; random search algorithm and, 142–44; species-area relationship and, 140, 145; stochastic simulation and, 142
organisms, 1, 7, 110, 114, 155
Otto, Sarah P., 10, 27, 97

parameters, 8
parametric bootstrapping, 122–24, 128
parametric statistics, 104
peer review, 1
phenomenological approach, 20, 27
physiological ecology, 1
plants: cellular automata and, 151–56; duckweed, 11–27, 35, 65–68; native, 151–56; shade and, 16–27
Plotkin, Joshua B., 114
Poisson distribution: generalized linear models and, 115–20; hypothesis testing and, 121–24; maximum likelihood, 105–9; model selection and, 110–14
populations: cellular automata and, 151–56; colonization and, 16 (see also colonization); difference equations and, 7–10; explaining data on, 101–28; extinction of, 81–87, 96, 132–38, 146–50; graphical thinking and, 132–39; growth of, 2, 14, 18–19, 22–23, 59, 61, 69, 133, 147, 150–51; Lotka-Volterra competition and, 22–27; maximum likelihood and, 103–9, 111, 113, 121, 124–25; metapopulations, 20, 81–86, 88,

90; modeling of, 1, 3, 10, 14, 17–28, 32, 57, 66–69, 73–75, 81–82, 88, 96, 102, 106, 108, 111–19, 123, 133–34, 146–47, 150–51, 155; multiple states and, 64–97; optimization and, 140–45; probability and, 1 (*see also* probability); samples from, 101–23; SIR models and, 28–32; stochastic simulation and, 146–50

population viability analyses, 146–47, 150

Possingham, Hugh P., 86–87

post-breeding census, 21

pre-breeding census, 21

predators, 57–60, 146

prediction: difference equations and, 7; exponential growth and, 11; Lotka-Volterra competition and, 23; model selection and, 110–11, 114; probability and, 35; SIR models and, 31; stochastic simulation and, 150; transition matrices and, 76

probability: absorption and, 81–86; base rate fallacy and, 55; basics of, 35–37; Bayesian, 3, 37, 50–56; coins and, 40; conditional, 47–55; contingency tables and, 43–52; diffusion and, 89–96; distribution, 57–61, 79 (*see also* distribution); duckweed and, 35; expected value, 41; frequentist, 36–38, 56, 79, 101–3, 121; generalized linear models and, 115–16, 120; hypothesis testing and, 126–28; independence and, 36, 39–42, 48–49, 58; interactions, 43; joint, 46–49, 53, 58; law of total, 54; log likelihood and, 106–9, 111–13, 118–19; marginal, 46, 48, 53–54; maximum likelihood, 103–9, 111, 113, 118–19, 121, 124–25; model selection and, 110, 112; P-value, 123, 126–27; random variables, 35–38, 41–47, 50, 53, 57–58; single-variable, 35, 38–42; statistics and, 3 (*see also* statistics); stochastic simulation and, 146–50; summation, 41–42; time steps and, 57, 61; transition matrices and, 79; two-variable, 43–49; uncertainty and, 2–3, 35–61

programming, 1–2

projection matrices, 70–71, 75, 77

projections: age-structured models and, 70–71, 75; cellular automata and, 152–53, 155; distribution and, 3, 57, 75; exponential growth and, 11, 13–14; initial conditions and, 8; linked, 70; logistic growth and, 16, 21; Lotka-Volterra competition and, 24; multi-state population models and, 3; SIR models and, 30–32; slope of, 10; stochastic simulation and, 147; time steps and, 9, 14, 21, 32, 57; transition matrices and, 77

prosecutor's fallacy, 55

P-value, 123, 126–27

quantitative ecology: basic principles of, 1–2; change over time, 5–32; difference equations and, 7; expanding the toolbox, 129–56; explaining data, 99–128; modeling multiple states, 63–97; online resources for, 4; rule-based models for, 155; uncertainty and, 33–61

quasi-extinction threshold, 146–50

rabbits, 57–61

rabies, 28–32

randomness: cellular automata and, 151; generalized linear models and, 118; hypothesis testing and, 125; independence and, 36; logistic growth and, 18, 20; maximum likelihood, 104; optimization and, 142; samples and, 102–4, 108, 112, 121–23; stochastic simulation and, 146, 150

random search algorithm, 142–44

random variables: Bayes's rule and, 37; independence and, 36, 39–42, 48–49, 58; probability and, 35–38, 41–47, 50, 53, 57–58; statistics and, 102, 104–5, 117–18, 121–22; stochastic simulation and, 147; transition matrices and, 79

reproductive factor, 14–15, 74
right eigenvector, 75
Rosenzweig, Michael L., 145
row stochastic matrices, 79–80, 87
Rule 126, 156

samples: bias and, 101–2; fitting the model and, 106; generalized linear models and, 115–20; hypothesis testing and, 121–28; log likelihood and, 106–9, 111–13, 118–19; maximum likelihood, 103–9, 111, 113, 118–19, 121, 124–25; model selection and, 110–14; random, 102–4, 108, 112, 121–23; statistics and, 101–23
Schluter, Dolph, 128
seeds: difference equations and, 7–10; generalized linear models and, 115; logistic growth and, 16–21; Lotka-Volterra competition and, 22–23, 26; maximum likelihood and, 103–9; model selection and, 110–14; transition matrices and, 79
shade: logistic growth and, 16–21; Lotka-Volterra competition and, 22–27
sigma, 41–42
single-variable probability, 35, 38–42
SIR models: difference equations and, 29–32; epidemics and, 38–41; Google Sheets and, 30; logistic growth and, 32; prediction and, 31; projections and, 30–32; rabies and, 28–32; time steps and, 28, 32
slope, 10
spatial ecology, 1
species-area relationship, 140, 145
stable age distribution, 74–75, 78
stationary state, 79
statistics, 2; basics of, 3, 101–2; blocking and, 42; data analysis and, 1; fitting the model and, 106; frequentist, 36–38, 56, 79, 101–3, 121; generalized linear models, 115–20; hypothesis testing, 121–28; log likelihood and, 106–9, 111–13, 118–19; maximum likelihood, 103–9, 111, 113, 118–19, 121, 124–25; model selection, 110–14; parametric, 104; random variables and, 102, 104–5, 117–18, 121–22; samples, 101–23; test statistic and, 122–28; uncertainty and, 37–38, 42, 54, 56, 61
stochastic simulation, 131; bears and, 146–50; birth-death chains and, 96; deterministic component of, 150; distribution and, 147–48; extinction and, 146–50; generalized linear models and, 117–18, 120; Google Sheets and, 148, 152; hypothesis testing and, 123; matrices and, 147; maximum likelihood and, 148; optimization and, 142; prediction and, 150; probability and, 146–50; projections and, 147; randomness and, 146–47, 150; uncertainty and, 150
Strauss, David, 109
succession, 76–80
summation, 41–42

test statistic, 122–28
ticks, 50–56
time steps: absorption and, 86; age-structured models and, 71–72; cellular automata and, 152–53; difference equations and, 7–10; diffusion and, 90–96; distribution and, 57, 61; exponential growth and, 12, 14–15; graphical thinking and, 133–38; linear equations and, 66, 72; logistic growth and, 21; Lotka-Volterra competition and, 24; matrices and, 65–66; probability and, 57, 61; projections and, 9, 14, 21, 32, 57; SIR models and, 28, 32
total probability, 42, 54
transition matrices: absorption and, 82, 85; difference equations and, 76–78; diffusion and, 88–97; distribution and, 78–79; forests and, 76–80; Google Sheets and, 78; prediction and, 76; projections and, 77; random variables and, 79; seeds and, 79; stationary state and, 78; succession and, 76–80; transition, 76–82, 85, 88–96

t-test, 120, 127–28
two-variable probability, 43–49

uncertainty: probability and, 35–37 (*see also* probability); statistics and, 37–38, 42, 54, 56, 61

vectors: absorption and, 85; age-structured models and, 71–75; diffusion and, 88–95; matrices and, 66–68, 71–79, 85, 88–95

Wald test, 128
Whitlock, Michael C., 128
Wilson, Edward O., 139
Wolfram, Stephen, 155–56

Zuur, Alain F., 120, 128